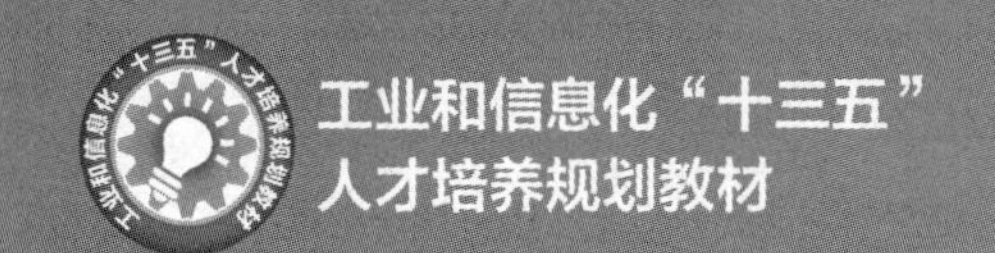

Java 高级程序设计|实战教程

Java Advanced Program Design

戴远泉 李超 秦争艳 ◎ 主编
董慧慧 麦琪琳 ◎ 副主编

人 民 邮 电 出 版 社
北 京

图书在版编目（CIP）数据

Java高级程序设计实战教程 / 戴远泉，李超，秦争艳主编. -- 北京 : 人民邮电出版社，2018.10（2019.7重印）
工业和信息化“十三五”人才培养规划教材
ISBN 978-7-115-48965-4

Ⅰ. ①J… Ⅱ. ①戴… ②李… ③秦… Ⅲ. ①JAVA语言－程序设计－高等学校－教材 Ⅳ. ①TP312.8

中国版本图书馆CIP数据核字(2018)第168368号

内 容 提 要

本书是在读者初步掌握Java的基础知识和技能之后，进一步学习Java高级编程的指导用书，主要内容包括Java编码规范、Java集合框架、Java反射机制、Java泛型机制、Java序列化机制、Java多线程机制、Java网络编程、Java数据库编程、综合实训等。本书能够帮助读者逐步领会Java的编程思想，并掌握Java的编程技能，为进一步学习J2EE框架技术奠定扎实的基础。

本书可作为应用型本科和高职高专院校计算机科学与技术、软件工程、软件技术等专业学生学习“Java高级程序设计”课程的教材及Java应用开发培训班的教材，也可作为Sun公司SCJP Java认证考试的辅导用书。

◆ 主　　编　戴远泉　李　超　秦争艳
　副 主 编　董慧慧　麦琪琳
　责任编辑　桑　珊
　责任印制　马振武

◆ 人民邮电出版社出版发行　　北京市丰台区成寿寺路11号
　邮编　100164　　电子邮件　315@ptpress.com.cn
　网址　http://www.ptpress.com.cn
　天津翔远印刷有限公司印刷

◆ 开本：787×1092　1/16
　印张：16.25　　　　　2018年10月第1版
　字数：451千字　　　　2019年7月天津第3次印刷

定价：46.00元

读者服务热线：(010)81055256　印装质量热线：(010)81055316
反盗版热线：(010)81055315
广告经营许可证：京东工商广登字20170147号

前言
Foreword

Java 是由 Sun Microsystems 公司于 1995 年 5 月推出的程序设计语言和 Java 平台的总称。

以 Sun 公司公布的 Java 框架结构为标准，Java 语言以 Java2 为中心，可分为以下 3 个组成部分。

（1）企业版 J2EE。该版本是以各大企业环境为中心而开发的一种以应用程序为主体的计算机网络平台，其中还包括 EJB、JSP 和 Servlet 3 个层次。

（2）标准版 J2SE。该版本中，Java 核心编程为图形用户界面的编程、工具包程序的编写以及数据库的程序编写。

（3）微型版 J2ME。该版本一直以消费品和各种嵌入式设备的网络应用平台为研究中心，主要涉及的领域为手机、手机中的各种无线游戏等，其核心技术为移动信息设备小程序。

Java 语言已是目前世界上流行的高级编程语言之一，正被广泛应用于计算机软件的开发，尤其是 Web 领域。自诞生以来，Java 迅速成为开发互联网应用程序的首选编程语言。

当前，应用型本科和高职高专院校开设的“Java 语言程序设计”课程相对应的教材主要讲述 Java 语言的基本语法（包括 Java 语言基础、数据类型、Java 类和对象），而高一级的软件工程专业普遍开设的 J2EE 课程对应的教材主要内容是 Servlet、JSP、SSH（Struts+Spring+Hibernate）及 SSM（Spring+SpringMVC+Mybatis）等企业级应用。从“Java 语言程序设计”到“J2EE 框架技术”等高级应用开发之间存在空白区，在课程开设及教学中，各门课程不能环环相扣，存在脱节现象。为了填补此空白区，本书在 Java 基础知识之上讲解了 Java 的高级技术和在实际 Java 项目的开发中所需的知识及其应用实例。在 Java 基础和 J2EE 应用之间起到了承前启后的作用。

本书特色如下。

（1）结构体系完整：本书体系完整，设计了 9 个应用领域，每个应用领域都是在实际软件开发中重要的或是频繁使用的知识点。

（2）实例源于真实：本书每个应用领域知识点对应的实例都源于或接近于真实项目，类的设计符合 Java 编程思想。

（3）讲解循序渐进：本书中对涉及的每个应用领域的知识点的讲解都由浅入深、循序渐进地展开。

（4）符合认知规律：本书采用“应用场景—相关知识—任务实施—拓展知识—拓展训练”的方式进行知识点的讲解，并配有课后小结、课后习题和上机实训。

本书内容如下。

本书设计了 9 个应用领域，每个应用领域的知识点都是在实际软件项目中得到大量应用的。

知识领域 1：Java 编码规范，讲解如何编写出符合规范、规则、惯例和模式的代码。

知识领域 2：Java 集合框架，讲解 List、Set 和 Map 等集合的使用。

知识领域 3：Java 反射机制，讲解 Java 反射机制的相关知识和应用。

知识领域 4：Java泛型机制，讲解泛型的相关知识及应用，包括泛型类、泛型方法和泛型接口。

知识领域 5：Java 序列化机制，讲解序列化的相关知识及应用，包括对象序列化和 JSON 序列化。

知识领域 6：Java 多线程机制，讲解多线程的相关知识及应用，主要包括线程的创建和启动、线程的生命周期、线程的管理等。

知识领域 7：Java 网络编程，讲解网络编程的相关知识及应用，包括基于 URL 读取网页内容、基于 TCP 编程和基于 UDP 编程。

知识领域 8：Java 数据库编程，讲解基于 Java 的数据库编程，包括数据库访问技术、JDBC 连接数据库、执行 SQL 操作。

知识领域 9：综合实训，讲述了一个完整的实训项目——“餐饮管理系统”，使用软件工程的思想进行需求分析、系统分析、系统设计、编码、测试等过程，完成本项目。

本书由戴远泉、李超、秦争艳任主编，董慧慧、麦琪琳任副主编，书中的每个例程代码都经过反复调试和测试。

由于编者水平有限，书中难免存在疏漏之处，欢迎广大读者批评指正。

编者

2018 年 5 月

目录
Contents

PART01

知识领域1

Java编码规范

知识目标

理解Java编码规范，包括：文件后缀名、Java源文件样式约定、注释规范、命名规范等。

■ 能力目标

1. 熟练阅读Java源码。
2. 熟练使用Java编码规则编写Java代码。

■ 素质目标

1. 培养查阅科技文档的能力。
2. 培养团队协作的能力。

1.1 应用场景

一个大型的软件项目是由一个团队来完成的，每个程序员在给包、类、变量、方法取名的时候，如果根本没有一点约定，只是随心所欲，可能会带来一系列问题。

编码规范是程序编码所要遵循的规则，保证代码的正确性、稳定性、可读性。规范编码有以下作用。

1. 规范的代码可以促进团队合作

一个项目大多都是由一个团队来完成的，统一的风格使得代码可读性大大提高，可以让开发人员尽快而彻底地理解新的代码，最大限度地提高团队开发的合作效率。

2. 规范的代码可以减少 BUG（漏洞）处理

没有对输入输出参数的规范，没有规范的异常处理，没有规范的日志处理等，不但会导致我们总是出现类似空指针这样低级的 BUG，而且还很难找到引起 BUG 的原因。相反，在规范的开发中，BUG 不但可以有效减少，查找 BUG 也变得轻而易举。

3. 规范的代码可以降低维护成本

好的编码规范可以尽可能地减少一个软件的维护成本，并且几乎没有软件能在其整个生命周期中均由最初的开发人员来维护。

4. 规范的代码有助于代码审查

代码审查可以及时纠正一些错误，可以对开发人员的代码规范做出监督。代码规范不仅使得开发统一，减少审查监督，而且让代码审查有据可查，大大提高了审查效率和效果，同时代码审查也有助于代码规范的实施。

5. 养成代码规范的习惯，有助于程序员自身的成长

长期的规范性编码还可以让开发人员养成好的编码习惯，甚至锻炼出更加严谨的思维习惯。规范的代码更有利于帮助开发人员理解开发语言、理解模式、理解架构，能够帮助开发人员快速提升开发水平。

因此，程序设计的标准化非常重要，原因在于这能提高开发团队各成员的代码的一致性，使代码更易理解，这意味着更易于开发和维护，从而降低了软件开发的总成本。为实现此目的，和其他语言类似，Java 语言也存在非强制性的编码规范。

1.2 相关知识

Java 常见编码规范包括：文件后缀名、源文件样式约定、注释规范、命名规范等。定义此种规范的目的在于让项目中所有的文档格式的统一，增加可读性。

1.2.1 文件后缀名

这部分列出了常用的文件类别及其后缀名，如表 1-1 所示。

表1-1 Java程序使用的文件后缀名

文件类别	文件后缀名
Java 源文件	.java
Java 字节码文件	.class

其中两者最本质的区别在于，.java 文件是供虚拟机运行时执行的文件，而 .class 文件可以让你在任何一台安装了 Java 虚拟机的机器上运行。

1.2.2 源文件样式约定

Java 源文件必须按顺序由以下 3 部分组成。

- 版权信息。
- 包和引入语句。
- 类和接口声明。

1. 版权信息

版权和版本信息必须在 Java 文件的开头，其他不需要出现在 Javadoc 的信息也可以包含在这里。

例如：

```
/**
*Title:  确定鼠标指针位置类
* Description: 确定鼠标指针当前在哪个作业栏位中并返回作业号
* @Copyright: Copyright (c) 2017
* @Company: daiinfo
* @author: daiyuanquan
* @version: 1.0
*/
```

2. 包和引入语句

package 行要在 import 行之前，import 中标准的包名要在本地的包名之前，而且按照字母顺序排列。如果 import 行中包含了同一个包中的不同子目录，则应该用 * 来处理。

例如：

```
package com.hbliti.net.stats;
import java.io.*;
import java.util.Observable;
import com.hbliti.util.Application;
```

3. 类和接口声明

每个 Java 源文件都包含一个单一的公共类或接口。类或接口的各部分代码顺序如下：

（1）常量声明；

（2）静态变量声明；

（3）成员变量声明；

（4）构造函数部分；

（5）Finalize 部分；

（6）成员方法部分；

（7）静态方法部分。

表 1-2 描述了类和接口声明的各个部分以及它们出现的先后次序。

表1-2 类和接口声明的各个部分顺序

序号	类/接口声明的各部分	说明
(1)	类 / 接口文档注释(/**……*/)	该注释中所需包含的信息
(2)	类或接口的声明	类头
(3)	类 / 接口实现的注释(/*……*/)// 如果有必要的话	该注释应包含任何有关整个类或接口的信息,而这些信息又不适合作为类 / 接口文档注释
(4)	常量	公共的、静态的、不可改变的,必须赋初始值(一旦赋值,不可改变)
(5)	类的静态成员变量	首先是类的公共变量,随后是保护变量,再后是包一级别的变量(没有访问修饰符),最后是私有变量
(6)	成员变量	首先是公共级别的,随后是保护级别的,再后是包一级别的(没有访问修饰符),最后是私有级别的
(7)	构造函数	应该用递增的方式写(比如:参数多的写在后面)
(8)	成员方法	这些方法应该按功能,而非作用域或访问权限分组。例如,一个私有的类方法可以置于两个公有的实例方法之间。其目的是为了更便于阅读和理解代码

(1)类 / 接口文档注释举例如下所示。

```
/**
* Title: 文件名称
* Description: 类内容的简介
*更新记录:
* 格式:[更新日期][修改的版本][操作人]内容
 * [2017-06-28][1.0][戴远泉]完善create方法。<br>
*
* Copyright: Copyright (c) 2017
* Company:   Daiinfo Co. Ltd.
* @version: 1.1
*/
```

(2)类或接口的声明举例如下所示。

```
public class CounterSet extends Observable implements Cloneable{
…
…
 }
```

(3)类 / 接口实现的注释。该注释应包含任何有关整个类或接口的信息,而这些信息又不适合作为类 / 接口文档注释。举例如下所示。

```
/**
* @see UserDao#save(User)
*/
public void save(User user) throws Exception{

}
```

(4)类的(静态)变量。静态变量是基本数据类型,这种情况下在类的外部不必创建该类的实例就可以直接使用。举例如下所示。

```
class Value{
  static int c=0;
  Value(){
    c=15;
  }
  Value(int i){
    c=i;
  }
  static void inc(){
    c++;
  }
}
class Count{
  public static void prt(String s){
    System.out.println(s);
  }
    Value v=new Value(10);
    static Value v1,v2;
    static{
      prt("v1.c="+v1.c+"  v2.c="+v2.c);
      v1=new Value(27);
      prt("v1.c="+v1.c+"  v2.c="+v2.c);
      v2=new Value(15);
      prt("v1.c="+v1.c+"  v2.c="+v2.c);
    }
  public static void main(String[] args){
    Count ct=new Count();
    prt("ct.c="+ct.v.c);
    prt("v1.c="+v1.c+"  v2.c="+v2.c);
    v1.inc();
    prt("v1.c="+v1.c+"  v2.c="+v2.c);
    prt("ct.c="+ct.v.c);
  }
}
```

（5）成员变量举例如下所示。

```
/**
* Packet counters
*/
protected int[] packets;
```

public 的成员变量必须生成文档（JavaDoc）。protected、private 和 package 定义的成员变量如果名字含义明确的话，可以没有注释。

（6）构造函数举例如下所示。

```
public CounterSet(){
  this.size = 100;
}
public CounterSet(int size){
  this.size = size;
}
```

应该用递增的方式写（比如参数多的写在后面）。

（7）成员方法举例如下所示。

```
/**
* param r1 - „„
* param r2 - „„
* param r3
* param r4
*/
protected final void setArray(int[] r1, int[] r2, int[] r3, int[] r4)  throws Illeg-
alArgumentException{
    // Ensure the arrays are of equal size
„„
}
```

（8）toString 方法举例如下所示。一般情况下，每一个类都应该定义 toString 方法。

```
public String toString() {
…
}
```

1.2.3　注释规范

代码注释是架起程序设计者与程序阅读者之间的通信桥梁，可以最大限度地提高团队开发合作效率，也是提高程序代码可维护性的重要环节之一。所以我们不是为写注释而写注释。

1. 注释编写的原则

（1）注释形式统一。在整个应用程序中，使用具有一致的标点和结构的样式来构造注释。如果在其他项目中发现它们的注释规范与这份文档不同，按照这份规范写代码，不要试图在既成的规范系统中引入新的规范。

（2）注释内容准确简洁。内容要简单、明了、含义准确，防止注释的多义性，错误的注释不但无益反而有害。

2. 注释类型的基本划分

（1）基本注释必须要添加，包括以下几种。

- 类（接口）的注释；
- 构造函数的注释；
- 方法的注释；
- 全局变量的注释；
- 字段 / 属性的注释；

简单的代码做简单注释，注释内容不大于 10 个字即可，另外，持久化对象或 VO 对象的 getter、setter 方法不需加注释。

（2）特殊必加的注释包括以下几种。

- 典型算法必须有注释；
- 在代码不明晰处必须有注释；
- 在代码修改处加上修改标识的注释；
- 在循环和逻辑分支组成的代码中加注释；
- 为他人提供的接口必须加详细注释。

具体的注释格式自行定义，要求注释内容准确简洁。

3. 注释的格式

（1）单行（single-line）注释格式为“//……”。

（2）块（block）注释格式为“/*……*/”。

（3）文档注释格式为“/**……*/”。

（4）Javadoc 注释标签语法如下：

- @author　对类的说明，标明开发该类模块的作者；
- @version　对类的说明，标明该类模块的版本；
- @see　对类、属性、方法的说明，参考转向，也就是相关主题；
- @param　对方法的说明，对方法中某参数的说明；
- @return　对方法的说明，对方法返回值的说明；
- @exception　对方法的说明，对方法可能抛出的异常进行说明。

例如：构造方法注释如下。

```
public class OkButton extends Button {
  /**
   * 构造方法的描述
   * @param name
   *   按钮上显示的文字
   */
  public Test(String name){
     ……
  }
}
```

1.2.4　命名规范

命名指系统中对包名、目录（类名）、方法、常量、变量等标识符的命名。标识符的命名力求做到统一、达意、简洁，遵循驼峰法则。

统一是指对于同一个概念，在程序中用同一种表示方法。例如对于供应商，既可以用 supplier，也可以用 provider，但是我们只能选定一个使用，至少在一个 Java 项目中保持统一。

达意是指标识符能准确地表达出它所代表的意义，如 newSupplier，OrderPaymentGatewayService 等；而 supplier1、service2、idtts 等则不是好的命名方式。

简洁是指，在统一和达意的前提下，用尽量少的标识符。如果不能达意，宁愿不要太简洁。例如，theOrderNameOfTheTargetSupplierWhichIsTransfered 太长，transferedTargetSupplierOrderName 则较好，但是 transTgtSplOrdNm 就不好了。省略元音的缩写方式不要使用，我们的英语往往还没有好到看得懂奇怪的缩写。

用驼峰法则是指单词之间不使用特殊符号分割，而是通过首字母大写来分割。例如推荐用 SupplierName，addNewContract，而不是 supplier_name，add_new_contract。

1. 包名命名规范

包名按如下规则组成。

[基本包].[项目名].[模块名].[子模块名]……

例如：com.czpost.eims

```
com.hepost.eims.until...
```

不得将类直接定义在基本包下，所有项目中的类、接口等都应当定义在各自的项目和模块包中。

2. 类名命名规范

（1）首字母大写。

类名要首字母大写，例如可以用 SupplierService，PaymentOrderAction；不要用 supplierService，paymentOrderAction。

（2）添加有含义的后缀。

类名往往用不同的后缀表达额外的意思，如表 1-3 所示。

表1-3　类名命名后缀名含义

后缀名	意义	举例
Service	表明这个类是个服务类，里面包含了给其他类提供业务服务的方法	PaymentOrderService
Impl	这个类是一个实现类，而不是接口	PaymentOrderServiceImpl
Inter	这个类是一个接口	LifeCycleInter
Dao	这个类封装了数据访问方法	PaymentOrderDao
Action	直接处理页面请求，管理页面逻辑的类	UpdateOrderListAction
Listener	响应某种事件的类	PaymentSuccessListener
Event	这个类代表了某种事件	PaymentSuccessEvent
Servlet	一个 Servlet	PaymentCallbackServlet
Factory	生成某种对象工厂的类	PaymentOrderFactory
Adapter	用来连接某种以前不被支持的对象的类	DatabaseLogAdapter
Job	某种按时间运行的任务	PaymentOrderCancelJob
Wrapper	这是一个包装类，为了给某个类提供没有的能力	SelectableOrderListWrapper
Bean	这是一个 POJO	MenuStateBean

3. 方法名命名规范

方法的命名规范有：首字母小写，如使用 addOrder()，不要用 AddOrder()；动词在前，如使用 addOrder()，不要用 orderAdd()。

动词前缀往往表达特定的含义，如表 1-4 所示。

表1-4　方法名命名后缀名含义

前缀名	意义	举例
create	创建	createOrder()
delete	删除	deleteOrder()
add	创建，暗示新创建的对象属于某个集合	addPaidOrder()
remove	删除	removeOrder()
init	初始化，暗示会做些诸如获取资源的特殊动作	initializeObjectPool

续表

前缀名	意义	举例
destroy	销毁，暗示会做些诸如释放资源的特殊动作	destroyObjectPool
open	打开	openConnection()
close	关闭	closeConnection()<
read	读取	readUserName()
write	写入	writeUserName()
get	获得	getName()
set	设置	setName()
prepare	准备	prepareOrderList()
copy	复制	copyCustomerList()
modity	修改	modifyActualTotalAmount()
calculate	数值计算	calculateCommission()
do	执行某个过程或流程	doOrderCancelJob()
dispatch	判断程序流程转向	dispatchUserRequest()
start	开始	startOrderProcessing()
stop	结束	stopOrderProcessing()
send	发送某个消息或事件	sendOrderPaidMessage()
receive	接受消息或时间	receiveOrderPaidMessgae()
respond	响应用户动作	responseOrderListItemClicked()
find	查找对象	findNewSupplier()
update	更新对象	updateCommission()

find方法在业务层尽量表达业务含义，例如使用 findUnsettledOrders()，表达查询未结算订单，而不要使用findOrdersByStatus()。数据访问层，find、update等方法可以实现对数据表的Select（查询）、Insert（插入）、Update（更新）、Delete（删除）等操作，如findByStatusAndSupplierIdOrderByName(Status.PAID, 345)。

4. 常量命名规范

常量必须为大写单词，下划线分隔的命名方式。常量一定是 static final 的字段，但是不是所有的 static final 字段都是常量，例如：

```
static final int NUMBER = 5;
static final ImmutableList<String> NAMES = ImmutableList.of("Ed", "Ann");
```

5. 变量命名规范

非常量的变量（类变量和实例成员变量）名必须采用小写单词驼峰命名方式（lowerCamelCase）。变量命名通常使用名词和名词短语，如 computedValue、index。

1.3　任务实施

任务　计算员工的月工资

1. 任务需求

某公司分为多个部门，每个部门有一个经理和多个员工，每个员工根据职称发基本工资。员工的工资由基本工资、日加班工资、日缺勤工资等组成。具体需求如下所示。

- 员工的基本信息，包括部门、职务、职称以及工资记录等信息。
- 能记录员工的每一个职称信息，并授予相应的职称，系统在计算员工工资的时候选取职称对应的最高职称津贴。

2. 任务分析

问题域中涉及多个类，包括职员类 Staffer、经理类 Manager、测试类 TestEmployee。

- Staffer 类：通过此类封装定义计算职员基本工资方法。
- Manager 类：通过此类封装定义计算经理基本工资方法。
- TestEmployee 类：调用方法并实现结果输出。

其类图关系如图 1-1 所示。

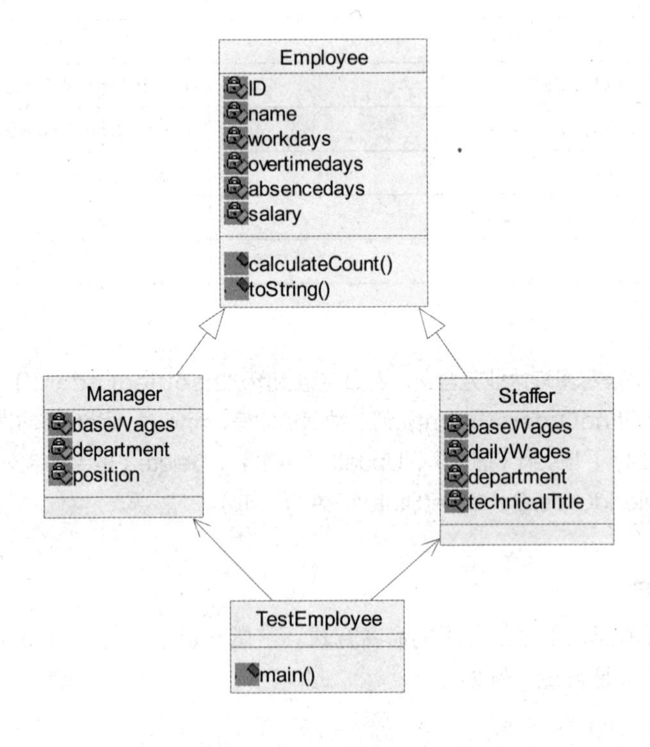

图1-1　类图关系

3. 任务实现

```
package com.daiinfo.seniorjava.ken1.implment;
/**
 *封装员工的信息和操作
```

```
 * @author daiyuanquan
 *
 */
public  class Employee {
    String ID;
    String name;

    int workdays;//工作天数
    int overtimedays;//加班天数
    int absencedays;//缺勤天数
    double salary;//月工资

    /**
     * 构造函数
     */
    public Employee(String ID) {
        // TODO Auto-generated constructor stub
        this.ID=ID;
    }

    /**
     * 构造函数
     */
    public Employee(String ID,String name) {
        // TODO Auto-generated constructor stub
        this.ID=ID;
        this.name=name;
    }


    /**
     * 计算员工的工资
     * @param workdays 工作天数
     * @param overtimedays 加班天数
     * @param absencedays 缺勤天数
     * @return 返回月工资总数
     */
    double calculateCount(int workdays,int overtimedays,int absencedays){
        double count=0.0;
        count=80.0*workdays+80*overtimedays-30*absencedays;
        return count;
    }

    /**
     * 转换字符串
     */
    public String toString(){
        return name+"\t"+salary;
    }
}
```

```
package com.daiinfo.seniorjava.ken1.implment;
public class Staff extends Employee{
    double bassewages=2000;
    double dailywages=50;//日工资
    String department;
    String technicaltitle;

    /**
     * 构造函数
     */
    public Staff(String ID,String name,String department,String technicaltitle){
        super(ID,name);
        this.department=department;
        this.technicaltitle=technicaltitle;
    }

    /**
     * 计算员工的工资
     * @param workdays 工作天数
     * @param overtimedays 加班天数
     * @param absencedays 缺勤天数
     * @return 返回月工资总数
     */
    double calculateCount(int workdays,int overtimedays,int absencedays){
        double count=0.0;
        count=bassewages+dailywages*overtimedays-dailywages*absencedays;
        return count;
    }

    /**
     * 转换字符串输出信息
     */
    public String toString(){
        return name+"\t"+salary;
    }
}
```

```
package com.daiinfo.seniorjava.ken1.implment;
public class Manager extends Employee {
    double basewages=3000;
    String department;//所在部门
    String positions;


    /**
     * 构造函数
     * @param ID
     * @param name
     * @param department
     */
    public Manager(String ID,String name,String department,String positions) {
```

```
        // TODO Auto-generated constructor stub
        super(ID, name);
        this.department=department;
        this.positions=positions;

    }

    /**
     * 计算经理的工资
     * @param workdays 工作天数
     * @param overtimedays 加班天数
     * @param absencedays 缺勤天数
     * @return 返回月工资总数
     */
    double calculateCount(int workdays,int overtimedays,int absencedays){
        double count=0.0;
        count=basewages+20*overtimedays-30*absencedays;
        return count;
    }
}
```

```
package com.daiinfo.seniorjava.ken1.implment;
/**
 * 测试类的引用
 *@Copyright: Copyright (c) 2017
 *@Company: daiinfo
 *
 *@author daiyuanquan
 *@version 1.0
 *@since 1.0
 */
public class TestEmployee {
    public static void main(String[] args) {
        // TODO Auto-generated method stub
        Manager manager=new Manager("001", "张三", "开发部","经理");
        double countsalary=manager.calculateCount(22, 3, 0);
        System.out.println(countsalary);

        Staff staff=new Staff("2001", "王好", "财务部", "会计师");
        double salary=staff.calculateCount(20, 5, 1);
        System.out.println(salary);
    }
}
```

运行结果如图 1-2 所示。

```
Markers  Properties  Servers  Data Source Explorer  Snippets  Console  Search
<terminated> TestEmployee [Java Application] C:\Java\jdk1.7.0_67\bin\javaw.exe (2017年7月13日 上午10:54:28)
3060.0
2200.0
```

图1-2　运行结果

1.4 拓展知识

相对于之前所提到的一些基本的编程需要注意的惯例，其他惯例（Miscellaneous Practices）也是非常重要的，下面就列出相对应的惯例。

1. 每次保存的时候，都让你的代码是最美的

使用 Ctrl+Shift+F 快捷键优化代码的格式和结构，调整代码。

2. 使用 log 而不是 System.out.println()

log 可以设定级别，可以控制输出到哪里，容易区分是在代码的什么地方打印的，而 System.out.print 则不行。而且，System.out.print 的速度很慢。所以，除非是有意的，否则都要用 log。至少在提交到 svn 之前把 System.out.print 换成 log。

3. 每个 if while for 等语句，都不要省略大括号 {}

```
if (a>b) {
    a++;
}
```

4. 善用 TODO

在代码中加入 //TODO：，大部分的 ide 都会帮你提示，让你知道你还有什么事没有做，比如：

```
if (order.isPaid()) {
   //TODO: 更新订单

}
```

5. 在需要留空的地方放一个空语句或注释

比如：

```
if (!exists(order)) {
   ;
}
```

或：

```
if (!exists(order)) {
     //nothing to do
}
```

6. 不要再对 boolean 值做 true、false 判断

比如：

```
if (order.isPaid() == true) {
   // Do something here
}
```

不如写成：

```
if (order.isPaid()) {
//Do something here
}
```

这样相对来说提高了代码的理解程度，更方便了其他人员阅读理解。

7. 减少代码嵌套层次

代码嵌套层次达 3 层以上时，一般人理解起来都会困难。下面的代码是一个简单的例子。

```
public void demo(int a, int b, int c) {
        if (a > b) {
            if (b > c) {
                doJobA();
            } else if (b < c) {
                doJobB();
            }
        } else {
            if (b > c) {
                if (a < c) {
                    doJobC();
                }
            }
        }
    }
```

减少嵌套的方法有很多，例如下面几个：

- 合并条件；
- 利用 return 以省略后面的 else ;
- 利用子方法。

比如上例，合并条件后成为

```
public void demo(int a, int b, int c) {
        if (a > b && b > c) {
            doJobA();
        }
        if (a > b && c > b) {
            doJobB();
        }
        if (a <= b && c < b && a < c) {
            doJobC();
        }
    }
```

如果利用 return 则成为

```
public void demo(int a, int b, int c) {
        if (a > b) {
            if (b > c) {
                doJobA();
                return;
            }
            doJobB();
            return;
        }
        if (b > c) {
            if (a < c) {
                doJobC();
```

```
            }
        }
    }
```

利用子方法，就是将嵌套的程序提取出来放到另外的方法里。

8. 程序职责单一

关注点分离是软件开发的真理。人类之所以能够完成复杂的工作，就是因为人类能够将工作分解到较小级别的任务上，在做每个任务时关注更少的东西。让程序单元的职责单一，可以使你在编写这段程序时关注更少的东西，从而降低难度，减少出错。

9. 变量的声明、初始化和被使用尽量放到一起

比方说如下代码：

```
int orderNum= getOrderNum();
//do something withou orderNum here
call(orderNum);
```

上例中的注释处代表了一段和 orderNum 不相关的代码。orderNum 的声明和初始化与被使用的地方相隔了其他的代码，这样做不好，不如这样：

```
//do something withou orderNum here
int orderNum= getOrderNum();
call(orderNum);
```

10. 缩小变量的作用域

能用局部变量的，不要使用实例变量；能用实例变量的，不要使用类变量。变量的生存期越短，它被误用的机会越小，同一时刻程序员要关注的变量的状态越少。实例变量和类变量默认都不是线程安全的，局部变量是线程安全的。比如如下代码：

```
public class OrderPayAction {
    private Order order;
    public void doAction() {
        order = orderDao.findOrder();
        doJob1();
        doJob2();
    }
    private void doJob1() {
        doSomething(order);
    }
    private void doJob2() {
        doOtherThing(order);
    }
}
```

上例中 order 只不过担当了在方法间传递参数之用，用下面的方法更好：

```
public class OrderPayAction {
    public void doAction() {
        order = orderDao.findOrder();
        doJob1(order);
        doJob2(order);
    }
```

```
    private void doJob1(Order order) {
        doSomething(order);
    }
    private void doJob2(Order order) {
        doOtherThing(order);
    }
}
```

11. 尽量不要用参数来带回方法运算结果

比如：

```
public void calculate(Order order) {
        int result = 0;
        // do lots of computing and store it in the result
        order.setResult(result);
    }
    public void action() {
        order = orderDao.findOrder();
        calculate(order);
        // do lots of things about order
    }
```

例子中 calculate 方法通过传入的 order 对象来存储结果，不如用如下方式写：

```
public int calculate(Order order) {
        int result = 0;
        // do lots of computing and store it in the result
        return result;
    }
    public void action() {
        order = orderDao.findOrder();
        order.setResult(calculate(order));
        // do lots of things about order
    }
```

1.5 拓展训练

良好的编码习惯可以大大提高可读性以及编码效率。请读者从一开始就养成一个良好的编码习惯。下面，我们通过以下校验邮件地址方法模板展示较为规范的编码格式。

```
package com.test;
/**
 * Title: 数组数据操作
 * Description: 演示一维数组和多维数组的初始化和基本操作
 * Filename: myArray.java
 */
 public class  MyArray{
   //初始化数组变量
   char[] cNum = {'1','2','3','4','5','6','7','8','9','0'};
   char[] cStr = {'a','b','c','d','e','f','g','h',
                  'i','j','k','l','m','n','o','p',
                  'q','r','s','t','u','v','w','x','y','z'};
   int[] iMonth = {31,28,31,30,31,30,31,31,30,31,30,31};
```

```
    String[] sMail = {"@","."};
  /**
   *<br>方法说明：校验电子邮件
   *<br>输入参数：String sPara 被校验的电子邮件字符
   *<br>返回类型：boolean 如果校验的格式符合电子邮件格式返回true，否则返回false
   */
    public boolean isMail(String sPara){
     for(int i=0;i<sMail.length;i++){
       if(sPara.indexOf(sMail[i])==-1)
         return false;
     }
     return true;
    }
  /**
   *<br>方法说明：判断是否是数字
   *<br>输入参数：String sPara。需要判断的字符串
   *<br>返回类型：boolean。如果都是数字类型，返回true，否则返回false
   */
    public boolean isNumber(String sPara){
      int iPLength = sPara.length();
      for(int i=0;i<iPLength;i++){
       char cTemp = sPara.charAt(i);
       boolean bTemp = false;
       for(int j=0;j<cNum.length;j++){
         if(cTemp==cNum[j]){
           bTemp = true;
           break;
         }
       }
       if(!bTemp) return false;
      }
     return true;
    }
  /**
   *<br>方法说明：判断是否都是英文字符
   *<br>输入参数：String sPara。要检查的字符
   *<br>返回类型：boolean。如果都是字符返回true，反之为false
   */
    public boolean isString(String sPara){
      int iPLength = sPara.length();
      for(int i=0;i<iPLength;i++){
       char cTemp = sPara.charAt(i);
       boolean bTemp = false;
       for(int j=0;j<cStr.length;j++){
         if(cTemp==cStr[j]){
           bTemp = true;
           break;
         }
       }
       if(!bTemp) return false;
      }
```

```
        return true;
    }
  /**
   *<br>方法说明：判断是否是闰年
   *<br>输入参数：int iPara。要判断的年份
   *<br>返回类型：boolean。如果是闰年返回true，否则返回false
   */
    public boolean chickDay(int iPara){
      return iPara%100==0&&iPara%4==0;
    }
  /**
   *<br>方法说明：检查日期格式是否正确
   *<br>输入参数：String sPara。要检查的日期字符
   *<br>返回类型：int。0为日期格式正确，-1为月或日不合要求，-2为年月日格式不正确
   */
    public int chickData(String sPara){
     @SuppressWarnings("unused")
     boolean bTemp = false;
     //所输入日期长度不正确
     if(sPara.length()!=10) return -2;
     //获取年
     String sYear = sPara.substring(0,4);
     //判断年是否为数字
     if(!isNumber(sYear)) return -2;
     //获取月份
     String sMonth = sPara.substring(5,7);
     //判断月份是否为数字
     if(!isNumber(sMonth)) return -2;
     //获取日
     String sDay = sPara.substring(8,10);
     //判断日是否为数字
     if(!isNumber(sDay)) return -2;
     //将年、月、日转换为数字
     int iYear = Integer.parseInt(sYear);
     int iMon = Integer.parseInt(sMonth);
     int iDay = Integer.parseInt(sDay);
     if(iMon>12) return -1;
     //闰年二月处理
     if(iMon==2&&chickDay(iYear)){
       if(iDay>29) return 2;
     }else{
       if(iDay>iMonth[iMon-1]) return -1;
     }
     return 0;
    }
  /**
   *<br>方法说明：主方法，测试用
   *<br>输入参数：
   *<br>返回类型：
   */
    public static void main(String[] arges){
```

```
        MyArray mA = new MyArray();
        //校验邮件地址
        boolean bMail = mA.isMail("tom@163.com");
        System.out.println("1 bMail is "+bMail);
        bMail = mA.isMail("tom@163com");
        System.out.println("2 bMail is "+bMail);
        //演示是否是数字
        boolean bIsNum = mA.isNumber("1234");
        System.out.println("1: bIsNum="+bIsNum);
        bIsNum = mA.isNumber("123r4");
        System.out.println("2: bIsNum="+bIsNum);
        //演示是否是英文字符
        boolean bIsStr = mA.isString("wer");
        System.out.println("1: bIsStr="+bIsStr);
        bIsStr = mA.isString("wer3");
        System.out.println("2: bIsStr="+bIsStr);
        //演示检查日期
        int iIsTime = mA.chickData("2003-12-98");
        System.out.println("1: iIsTime="+iIsTime);
        iIsTime = mA.chickData("2003-111-08");
        System.out.println("2: iIsTime="+iIsTime);
        iIsTime = mA.chickData("2003-10-08");
        System.out.println("3: iIsTime="+iIsTime);
        iIsTime = mA.chickData("2000-02-30");
        System.out.println("4: iIsTime="+iIsTime);
    }
}
```

运行结果如图 1-3 所示。

```
Markers  Properties  Servers  Data Source Explorer  Snippets  Console  Search
<terminated> MyArray [Java Application] C:\Java\jdk1.7.0_67\bin\javaw.exe (2017年6月26日 上午1:50:41)
1 bMail is true
2 bMail is false
1: bIsNum=true
2: bIsNum=false
1: bIsStr=true
2: bIsStr=false
1: iIsTime=-1
2: iIsTime=-2
3: iIsTime=0
4: iIsTime=2
```

图1-3　运行结果

1.6　课后小结

1. 规范需要在平时编码过程中注意，是一个慢慢养成的好习惯。
2. 命名规范：所有的标识符只能用英文字母、数字、下划线，并采用驼峰标识。
3. 注释规范：注释有单行注释 “//”，多行注释 “/*……*/”，文档注释 “/**……*/”。
4. 排版规范：代码应避免一行长度超过 60 个字符。
5. 声明规范：一行声明一个变量，只在代码块的开始处声明变量。
6. 语句规范：一条语句占用一行。

7. 编程规范。用于设置的方法前缀必须是 set；用于检索一个布尔类型的方法前缀必须是 is，而用于检索其他方法前缀必须是 get。程序中应尽可能少使用数字（或字符），尽可能定义静态变量来说明该数字（或字符）的含义。

1.7 课后习题

一、填空题

1. 相对独立的程序块之间、变量说明之后必须加________。

2. 类的注释部分，描述部分说明该类或者接口的功能、作用、使用方法和注意事项，每次修改后增加作者、新版本号和当天的日期，@since ________，@deprecated 表示________。

3. 比较操作符，赋值操作符“=”、“+=”，算术操作符“+”、“%”，逻辑操作符“&&”、“&”，位域操作符“<<”、“^”等双目操作符的前后加________。

4. Java 中的注释有 3 种形式：文档注释、多行注释和________。

5. 方法的文档中，块标记 @param 用于说明________的含义，@return 用于说明________含义。

二、选择题

1. 下列使用异常的做法错误的是________。

A. 在程序中使用异常处理还是使用错误返回码处理，根据是否有利于程序结构来确定，并且异常和错误码不应该混合使用，推荐使用异常

B. 一个方法不应抛出太多类型的异常。throws/exception 子句标明的异常最好不要超过 3 个

C. 异常捕获尽量不要直接使用 catch(Exception ex)，应该把异常细分处理

D. 程序内抛出的异常本身就可说明异常的类型、抛出条件，可不填写详细的描述信息。捕获异常后用 exception.toString() 取到详细信息后保存

2. 下列说法错误的是________。

A. 段代码各语句之间有实质性关联并且是完成同一件功能的，那么可考虑把此段代码构造成一个新的方法

B. 源程序中关系较为紧密的代码应尽可能相邻

C. 程序中可同时使用错误码和异常进行处理，推荐使用异常

D. 方法参数建议不超过 5 个

3. 下面对类、方法、属性的说法不符合编程规范的有________。

A. 不要覆盖父类的私有方法

B. 类中不要使用非私有的非静态属性

C. 类定义

```
{
    类的私有属性定义
    类的公有属性定义
    类的保护属性定义
    类的私有方法定义
    类的公有方法定义
    类的保护方法定义
}
```

D. 类私有方法的最大规模建议为 15 个

4. 排版时，代码缩进应该采用的方式是________。

A. Tab 缩进

B. 2 个空格缩进

C. 4 个空格缩进

D. 8 个空格缩进

5. 下列关于注释的说法正确的是________。

A. 包注释可有可无，一般大家都是看类注释和方法注释

B. 可以把一个类的类注释改为它的文件注释

C. 类注释应该放在 package 关键字之后，class 或者 interface 关键字之前

D. 文件注释应该使用 Javadoc 定义的方式注释，保证能够被收集并形成 doc 文档

三、简答题

1. 请简述类编写规范。
2. 请简述 Java 类中方法的编写规范。
3. 请简述合适的命名对提高代码质量的价值。
4. 请简述 Java 的命名规则。

1.8 上机实训

本例计算几何图形的面积、周长，创建抽象类 Shape，圆类 Circle、矩形类 Rectangle 继承抽象类 Shape。测试类 TestShape 完成各种图形的测试及其方法调用，计算其面积和周长。类图如图 1-4 所示。

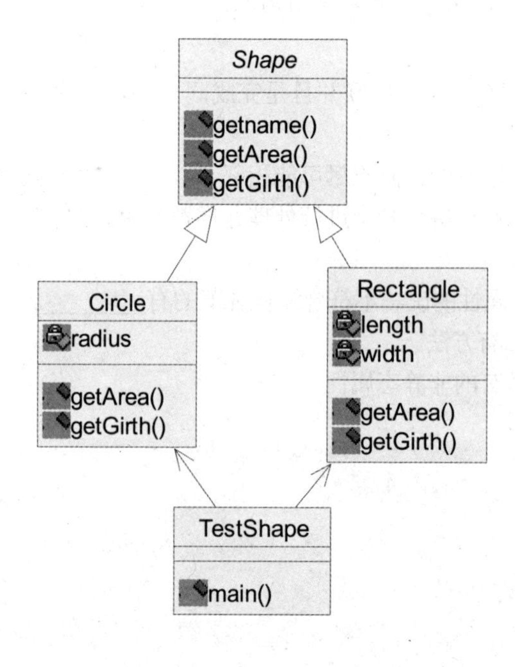

图1-4 图形类图

实现代码如下所示。

```
package com.daiinfo.seniorjava.ken1.training;
public abstract class Shape {

    /**
     * 获得图形的名称
     * @return 返回图形的名称
     */
    public String getName() {
        return this.getClass().getSimpleName();
    }

    /**
    * 获得图形的面积
    * @return 返回图形的面积
    */
    public abstract double getArea();

    /**
     * 获得图形的周长
     * @return 返回图形的周长
     */
    public abstract double getGirth();
}
```

```
package com.daiinfo.seniorjava.ken1.training;
public class Circle extends Shape {
    private double radius;

    /**
     * 构造函数  构造一个圆
     * @param radius为圆的半径
     */
    public Circle(double radius) {

    }

    /**
     * 计算圆的面积
     * @return 返回圆的面积
     */
    @Override
    public double getArea() {
        return Math.PI * Math.pow(radius, 2);
    }
    /**
     * 计算圆的周长
     * @return 返回圆的周长
     */
```

```
    @Override
    public double getGirth() {
        // TODO Auto-generated method stub

    }
}
```

```
package com.daiinfo.seniorjava.ken1.training;
public class Rectangle extends Shape {
    private double length;
    private double width;

    /**
     * 构造函数 构造一个矩形
     * @param length
     * @param width
     */
    public Rectangle(double length, double width) {//获得矩形的长和宽

    }

    /**
     * 获得矩形的面积
     */
    @Override
    public double getArea() {//计算矩形的面积
        // TODO Auto-generated method stub
    }

    /**
     * 获得矩形的周长
     */
    @Override
    public double getGirth() {
        // TODO Auto-generated method stub

    }
}
```

```
package com.daiinfo.seniorjava.ken1.training;
public class TestShape {
    public static void main(String[] args) {
        // TODO Auto-generated method stub

    }
}
```

PART 02

知识领域2

Java集合框架

知识目标

1. 理解Java的集合框架体系。
2. 掌握List、Set、Map 接口的具体实现、内部结构、特殊的方法和适用场景等。

■ 能力目标

1. 熟练使用List、Set、Map 接口的具体实现类完成对集合的“增删改查”操作。
2. 根据各个不同的场景选择适合的框架来分析问题和解决问题。

■ 素质目标

1. 能够阅读科技文档和撰写分析文档。
2. 能够查阅jdk API。

2.1 应用场景

在 Java 编程时，常常需要集中存放多个数据。当然我们可以使用数组来保存多个对象。但数组长度不可变化；而且数组无法保存具有映射关系的数据。

为了保存数量不确定的数据，以及保存具有映射关系的数据（也被称为关联数组），就要使用 Java 提供的集合类。JDK 提供了大量优秀的集合实现供开发者使用，合格的程序员必须要能够通过功能场景和性能需求选用最合适的集合，这就要求开发者必须熟悉 Java 的常用集合框架类。

2.2 相关知识

在现实生活中，集合被理解为很多事物凑在一起；在数学中，集合为具有共同属性的事物的总体。通常情况下，我们把具有相同性质的一类东西，汇聚成一个整体，就可以称为集合。通常集合有两种表示法，一种是列举法，比如集合 A={1,2,3,4}；另一种是性质描述法，比如集合 B={X|0<X<100 且 X 属于整数}。集合论的奠基人康托尔在创建集合理论时给出了许多公理和性质，这都成为后来集合在其他领域应用的基础。

2.2.1 集合框架

集合是存放数据的容器；框架是类库的集合。集合框架就是为表示和操作集合而规定的一种统一的、标准的体系结构。

任何集合框架都包含三大块内容：对外的接口、接口的实现类和对集合运算的算法。我们可以把一个集合看成一个微型数据库，操作包括“增、删、改、查”4 种。

2.2.2 Java 集合框架

在 Java 语言中，Java 语言的设计者对常用的数据结构和算法做了一些规范（接口）和实现（具体实现接口的类）。所有抽象出来的数据结构和操作（算法）统称为 Java 集合框架（JavaCollectionFramework），如图 2-1 所示。

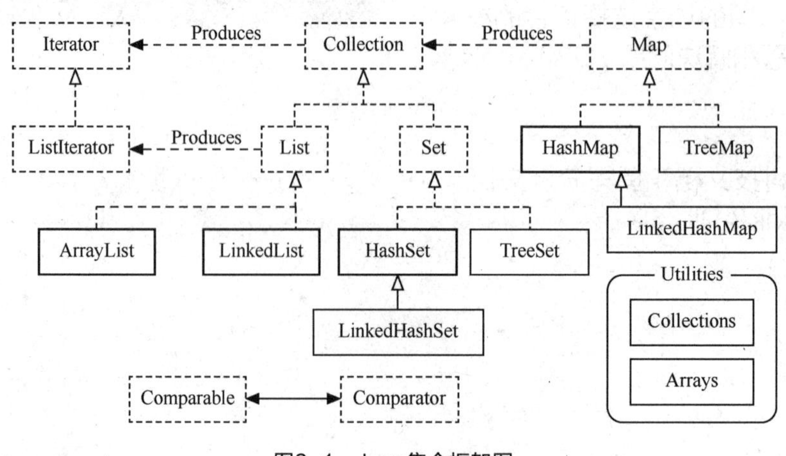

图2-1　Java集合框架图

我们经常用到的 Collection、List、Set、Queue 和 Map 都是接口（Interface），不是具体的类实现。Java 集合框架支持 3 种类型的集合：规则集（Set），线性表（List）和图（Map），如表 2-1 和图 2-2 所示。

表2-1　集合类和接口比较表

集合类和接口		是否有序	元素是否重复
Collection		否	是
List		是	是
Set	AbstractSet	否	否
	HashSet	否	否
	TreeSet	是（用二叉排序树）	否
Map	AbstractMap	否	使用 key-value 来映射和存储数据，key 必须唯一，value 可以重复
	HashMap	否	
	TreeMap	是（用二叉树）	

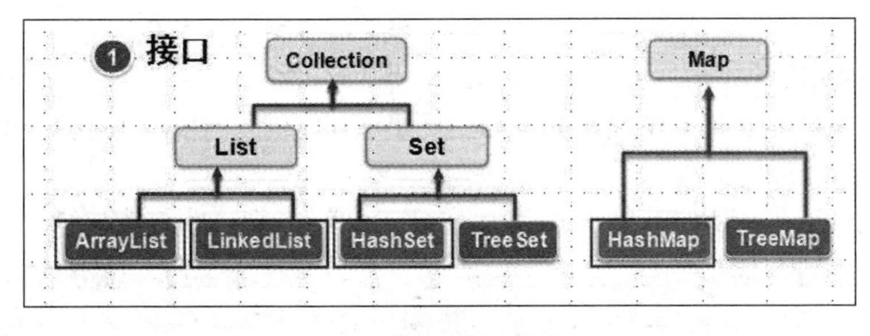

图2-2　集合类接口关系图

2.2.3　Java 集合框架的使用

Java 程序员在具体应用时，不必考虑数据结构和算法的实现细节，只需要用这些类创建出来一些对象，然后直接应用就可以了，这样就大大提高了编程效率。

2.3　任务实施

任务一　使用 List 存取用户信息，并做增删改查操作

1. 任务知识

（1）List 接口

List 继承自 Collection 接口。List 是一种有序集合，List 中的元素可以根据索引（顺序号：元素在集合中处于的位置信息）进行查询 / 删除 / 插入操作。

跟 Set 集合不同的是，List 允许有重复元素。对于满足 e1.equals(e2) 条件的 e1 与 e2 对象元素，可以同时存在于 List 集合中。

（2）List 实现类

List 接口的实现类主要有 ArrayList、LinkedList、Vector、Stack 等。

（3）ArrayList 常用方法

一般主要使用的 ArrayList 方法，如表 2-2 所示。

表2-2　ArrayList主要方法

返回值类型	方法名	说明
boolean	add(E e)	将指定的元素添加到此列表的尾部
void	add(int index，E element)	将指定的元素插入此列表中的指定位置
Object	remove(int index)	移除此列表中指定位置上的元素
boolean	remove(Object o)	移除此列表中首次出现的指定元素（如果存在）
Object	set(int index,Object obj)	用指定的元素替代此列表中指定位置上的元素
Object	get(int index)	返回此列表中指定位置上的元素
int	indexOf(Object obj)	返回此列表中首次出现的指定元素的索引，或如果此列表不包含元素，则返回 -1
int	lastIndexOf(Object obj)	返回此列表中最后一次出现的指定元素的索引，或如果此列表不包含索引，则返回 -1

（4）List 一般用法

在使用 List 集合时，通常情况下声明为 List 类型，实例化时根据实际情况的需要，实例化为 ArrayList 或 LinkedList。

① 声明一个 list 的示例如下：

```
List<String>  l1 = new ArrayList<String>();     //利用ArrayList类实例化List集合
List<String>  l2 = new LinkedList<String>();    //利用LinkedList类实例化List集合
```

② 向 list 中存值的示例如下：

```
l1.add("张三");
l2.set(1, "汪涵");     //将索引位置为1的对象e修改为对象“汪涵”
```

③ 遍历 list 的示例如下：

```
Iterator it = l2.iterator();
while (it.hasNext()) {
     System.out.println(it.next());
  }
}
```

2. 任务需求

使用 List 存取用户信息，并做增删改查操作。

3. 任务分析

一个班级之中存在若干个学生，通过一个实体类定义一个学生相对应的基本信息，然后通过一个 List 集合进行存储，实现对学生基本信息的 CURD。类图如图 2-3 所示。

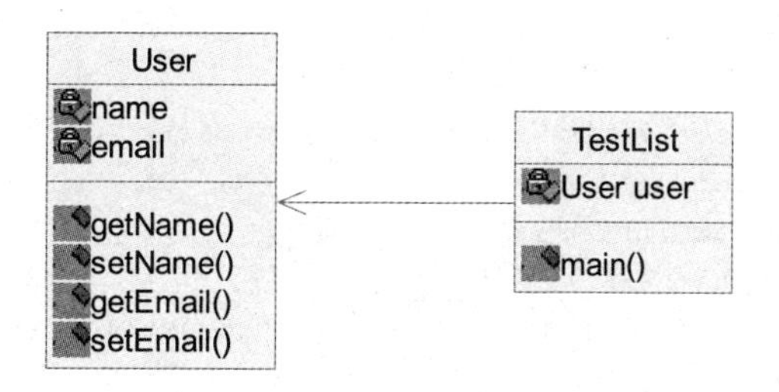

图2-3　测试存取用户信息的类图

4. 任务实现

User.java

```
package com.daiinfo.seniorjava.ken2.implement.list;
public class User {
    String name;
    String email;
    public User(String name, String emailString) {
        this.name = name;
        this.email = emailString;
    }
    public String getName() {
        return name;
    }
    public void setName(String name) {
        this.name = name;
    }
    public String getEmail() {
        return email;
    }
    public void setEmail(String email) {
        this.email = email;
    }
}
```

TestList.java

```
package com.daiinfo.seniorjava.ken2.implement.list;
import java.util.ArrayList;
import java.util.Iterator;
import java.util.List;
import java.util.ListIterator;
public class TestList {
    public static void main(String[] args) {
        List<User> list=new  ArrayList<User>();
        list.add(new User("小明", "xiaoming@qq.com"));

        User xiaozhang=new User("小张","xiaozhang@qq.com");
        User xiaohu=new User("小胡", "xiaohu@qq.com");
        list.add(xiaozhang);
        list.add(xiaohu);
```

```
        //遍历
        //方法1
        ListIterator<User> it1 = list.listIterator();
        System.out.println("方法1 ");
        while(it1.hasNext()){
            User user=it1.next();
            System.out.println(user.getName()+user.getEmail());
        }
      //方法2
        System.out.println("方法2 ");
        for(Iterator<User> it2 = list.iterator();it2.hasNext();){
             User user=it2.next();
             System.out.println(user.getName()+user.getEmail());
        }

      //方法3
        System.out.println("方法3 ");
        for(User tmp:list){
             System.out.println(tmp.getName()+tmp.getEmail());
        }
        //方法4
        System.out.println("方法4 ");
        for(int i = 0;i < list.size(); i ++){
             System.out.println(list.get(i).getName()+list.get(i).getEmail());
        }

        //移除列表中的元素
        list.remove(1);

        System.out.println("方法3 ");
        for(User tmp:list){
             System.out.println(tmp.getName()+tmp.getEmail());
        }
    }
}
```

运行结果如图 2-4 所示。

```
Markers  Properties  Servers  Data Source Explorer  Snippets  Console  Search
<terminated> TestList [Java Application] C:\Java\jdk1.7.0_67\bin\javaw.exe (2017年7月14日 下午5:50:42)
方法1
小明xiaoming@qq.com
小张xiaozhang@qq.com
小胡xiaohu@qq.com
方法2
小明xiaoming@qq.com
小张xiaozhang@qq.com
小胡xiaohu@qq.com
方法3
小明xiaoming@qq.com
小张xiaozhang@qq.com
小胡xiaohu@qq.com
方法4
小明xiaoming@qq.com
小张xiaozhang@qq.com
小胡xiaohu@qq.com
方法3
小明xiaoming@qq.com
小胡xiaohu@qq.com
```

图2-4　ArrayList测试运行结果图

说明：

① 使用 add() 方法向列表的尾部或指定位置添加指定的元素。

```
list.add(new User("小明", "xiaoming@qq.com"));
User xiaozhang=new User("小张","xiaozhang@qq.com");
list.add(xiaozhang);
```

② 使用 set() 方法替换列表中指定位置的元素；使用 get() 方法返回列表中指定位置的元素。

```
list.get(i);
list.set(1,"小胡");
```

③ 使用 remove(int index) 方法移除列表中指定位置的元素。

④ 使用 listIterator() 方法返回此列表元素的列表迭代器。利用迭代最大的好处就程序员不用再去为了索引越界等等异常所苦恼了，只需在迭代过程中对列表元素进行操作即可。

```
ListIterator<Integer> iterator = list.listIterator();
while (iterator.hasNext()) {
    System.out.println();
}
```

任务二　使用 Set 存取数据，并做增删改查操作

1. 任务知识

（1）Set 接口

Set 是继承于 Collection 的接口。Java 中的 set 接口有如下特点。

- 不允许出现重复元素；
- 集合中的元素位置无顺序；
- 有且只有一个值为 null 的元素。

（2）Set 接口的实现类

实现了 Set 接口的主要有 HashSet、TreeSet、LinkedHashSet。

HashSet 依赖于 HashMap，它实际上是通过 HashMap 实现的。HashSet 中的元素是无序的。

TreeSet 依赖于 TreeMap，它实际上是通过 TreeMap 实现的。TreeSet 中的元素是有序的。

（3）HashSet 常用方法

HashSet 是存在于 Java.util 包中的类。同时也被称为集合，一般常用的方法如表 2-3 所示。

表2-3　HashSet常用方法

返回值类型	方法名	说明
boolean	add(E e)	如果此 set 中尚未包含指定元素，则添加指定元素
void	clear()	从此 set 中移除所有元素
boolean	contains(Object o)	如果此 set 包含指定元素，则返回 true
boolean	isEmpty()	如果此 set 不包含任何元素，则返回 true
Iterator<E>	iterator()	返回对此 set 中元素进行迭代的迭代器
boolean	remove(Object o)	如果指定元素存在于此 set 中，则将其移除
int	size()	返回此 set 中的元素的数量（set 的容量）

（4）Set 接口的一般用法。

在使用 Set 集合时，通常情况下声明为 Set 类型，实例化时根据实际情况的需要，实例化为 HashSet 或 TreeSet。

① 创建 HashSet 对象的示例如下：

```
HashSet hashSet=new HashSet();
```

② 添加元素的示例如下：

```
hashset.add("abc");
```

③ 删除元素的示例如下：

```
hashset.clear()                //从此 set 中移除所有元素。
hashset.remove(Object o)       //如果指定元素存在于此 set 中，则将其移除。
```

④ 遍历 HashSet 的示例如下：

```
Iterator it = hashset.iterator();
            while(it.hasNext()){}
```

2. 任务需求

使用 Set 存储学生信息，并进行 CRUD 操作。

3. 任务分析

一个班级之中存在若干个学生，通过一个实体类定义一个学生相对应的基本信息，然后通过一个 Set 集合进行存储，实现对学生基本信息的 CURD。

类图关系如图 2-5 所示。

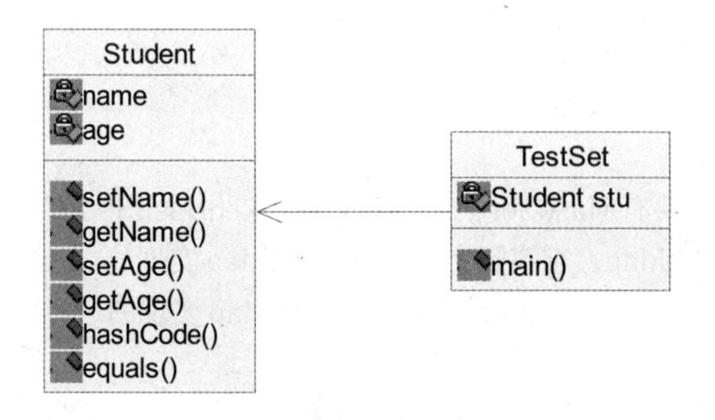

图2-5　HashSet测试类图

4. 任务实现

Student.java

```
package com.daiinfo.seniorjava.ken2.implement.hashset;
class Student {
    private String name;
    private int age;
    public Student(String name, int age) {
        this.name = name;
        this.age = age;
    }
    // 复写hashCode方法
```

```
    @Override
    public int hashCode() {
        return 60;
    }
    // 复写equals方法
    @Override
    public boolean equals(Object arg0) {
        if (!(arg0 instanceof Student)) {
            return false;
        }
        Student studet = (Student) arg0;
        return this.name.equals(studet.name) && this.age == studet.age;
    }
    public String getName() {
        return name;
    }
    public void setName(String name) {
        this.name = name;
    }
    public int getAge() {
        return age;
    }
    public void setAge(int age) {
        this.age = age;
    }
}
```

TestHashSet.java

```
package com.daiinfo.seniorjava.ken2.implement.hashset;
import java.util.HashMap;
import java.util.HashSet;
import java.util.Iterator;
import java.util.Map.Entry;
import java.util.Set;
public class TestHashSet{
    public static void hashSet1() {
        HashSet<String> hashSet = new HashSet<String>();
        hashSet.add("java001");
        hashSet.add("java01");
        hashSet.add("java011");
        hashSet.add("java002");
        hashSet.add("java004");
        // 使用常用迭代器获取输出内容
        Iterator<String> iterator = hashSet.iterator();
        while (iterator.hasNext()) {
            String next = iterator.next();
            System.out.println(next);
        }
    }
    public static void hashSet2() {
        HashSet<Student> hashSet = new HashSet<Student>();
```

```
        hashSet.add(new Student("zhangsan1", 21));
        hashSet.add(new Student("zhangsan1", 21));
        hashSet.add(new Student("zhangsan2", 21));
        hashSet.add(new Student("zhangsan3", 23));
        hashSet.add(new Student("zhangsan4", 24));
        // 使用常用迭代器获取值
        Iterator<Student> iterator = hashSet.iterator();
        while (iterator.hasNext()) {
            Student next = (Student) iterator.next();
            System.out.println(next.getName() + " " + next.getAge());
        }
    }
    public static void main(String[] args) {
        hashSet1();
        hashSet2();
    }
}
```

运行结果如图 2-6 所示。

```
Markers  Properties  Servers  Data Source Explorer  Snippets  Console  Search
<terminated> TestHashSet [Java Application] C:\Java\jdk1.7.0_67\bin\javaw.exe (2017年7月14日 下午6:29:48)
java001
java011
java01
java004
java002
zhangsan4 24
zhangsan3 23
zhangsan2 21
zhangsan1 21
```

图2-6　HashSet测试运行结果图

任务三　使用 Map 存取数据，并做增删改查操作

1. 任务知识

（1）Map 接口

Map 提供了一种映射关系，其中的元素是以键值对（key-value）的形式存储的，能够实现根据 key 快速查找 value。Map 中的键值对以 Entry 类型的对象实例形式存在。键值（key 值）不可重复，value 值可以重复，一个 value 值可以和很多 key 值形成对应关系，每个键最多只能映射到一个值。Map 支持泛型，形式如：Map<K,V>。Map 中使用 put(K key, V value) 方法添加。

（2）已知实现类

在 Java.util 包中接口 Map<K,V> 存储键值对，作为一个元组存入。元组以键作为标记，键相同时，值覆盖。

类型参数有：

- K——此映射所维护的键的类型；
- V——映射值的类型。

其已知实现类为 HashMap、TreeMap。

（3）HashMap 常用操作说明

HashMap 是一个散列表，它存储的内容是键值对（key-value）映射。HashMap 继承于 AbstractMap，实现了 Map、Cloneable、Java.io.Serializable 接口。HashMap 的实现不是同步的，这意味着它不是线程安全的。它的 key、value 都可以为 null。

HashMap 的一般主要方法如表 2-4 所示。

表2-4　HashMap接口主要方法

返回值类型	方法名	说明
V	put(K key, V value)	在此映射中关联指定值与指定键
V	get(Object key)	返回指定键所映射的值；如果对于该键来说，此映射不包含任何映射关系，则返回 null
Set<K>	keySet()	返回此映射中所包含的键的 Set 视图
Collection<V>	values()	返回此映射所包含的值的 Collection 视图
V	remove(Object key)	从此映射中移除指定键的映射关系（如果存在）
Boolean	containsValue(Object value)	如果此映射将一个或多个键映射到指定值，则返回 true
Boolean	containsKey(Object key)	如果此映射包含对于指定键的映射关系，则返回 true
Boolean	isEmpty()	如果此映射不包含键 - 值映射关系，则返回 true
int	size()	返回此映射中的键 - 值映射关系数

（4）Map 的一般用法

① 声明一个 Map 的示例如下：

```
Map map = new HashMap();
```

② 向 map 中存值的示例如下（注意 map 是以 key-value 的形式存放的）：

```
map.put("sa","dd");
```

③ 从 map 中取值的示例如下：

```
String str = map.get("sa").toString,
```

结果是：

```
str = "dd'
```

④ 遍历一个 map，从中取得 key 和 value 的示例如下：

```
Map map = new HashMap();
for (Object obj : map.keySet()) {
    Object value = map.get(obj);
}
```

2. 任务需求

使用 Map 存储学生信息，并进行 CRUD 操作。

3. 任务分析

一个班级之中存在若干个学生，通过一个实体类定义一个学生相对应的基本信息，然后通过一个 Map 集合进行 key-value 键值存储，实现对学生基本信息的 CURD。类图如图 2-7 所示。

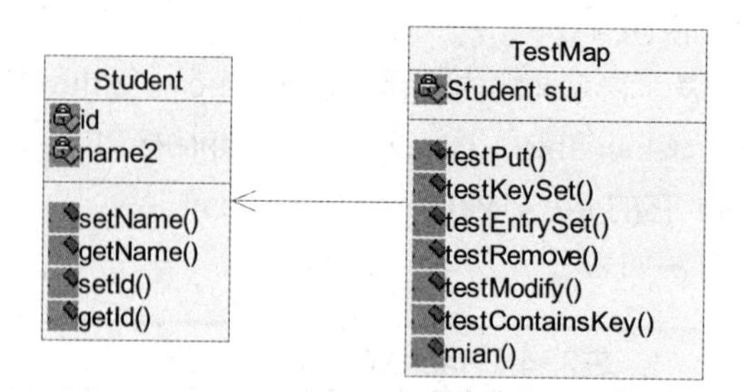

图2-7　HashMap测试类图

4. 任务实现

Student.java

```
package  com.daiinfo.seniorjava.ken2.implement.hashmap;
import java.util.HashSet;
import java.util.Set;
/**
 * 学生类
 * @author daiyuanquan
 * @version 1.0
 */
public class Student {
    public String id;
    public String name;

    /**
     * 构造函数
     * @param id
     * @param name
     */
    public Student(String id,String name){
        this.id = id;
        this.name = name;
    }
}
```

TestMap.java

```
package com.daiinfo.seniorjava.ken2.implement.hashmap;
import java.util.HashMap;
import java.util.Map;
import java.util.Map.Entry;
import java.util.Scanner;
import java.util.Set;
/**
 * 测试Map应用
 * @author daiyuanquan
 *
 */
public class TestMap {
    //创建一个Map属性用来承装学生对象
    public Map<String,Student> student;
```

```
/*
 * 在构造器中初始化学生属性
 */
public TestMap(){
    this.student = new HashMap<String,Student>();
}

/*
 * 添加方法：输入学生ID，判断是否被占用，
 * 如果未被占用，则输入姓名，创建新学生对象，添加到student中
 */
public void testPut(){
    //创建一个Scanner对象，输入学生ID
    Scanner sc = new Scanner(System.in);
    int i = 0;
    while(i<3){
        System.out.println("请输入学生ID：");
        String stuID = sc.next();
        Student stu = student.get(stuID);
        if(stu == null){
            System.out.println("请输入学生姓名：");
            String stuName = sc.next();
            Student newStudent = new Student(stuID,stuName);
            student.put(stuID, newStudent);
            System.out.println("成功添加学生："+student.get(stuID).name);
            i++;
        }else{
            System.out.println("该学生ID已被占用！");
            continue;
        }
    }
}

/*
 * 测试Map的keySet方法
 */
public void testKeySet(){
    //通过keySet方法，返回Map中所有“键”的Set集合
    Set<String> keySet = student.keySet();
    //取得student的容量
    System.out.println("总共有"+student.size()+"个学生；");
    //遍历keySet，取得每一个键，再调用get方法取得每个键对应的value
    for (String stuID : keySet) {
        Student stu = student.get(stuID);
        if(stu != null){
            System.out.println("学生："+stu.name);
        }
    }
}
```

```
    /*
     * 通过entrySet来遍历Map
     */
    public void testEntrySet(){
        //通过entrySet返回Map中所有的键值对
        Set<Entry<String,Student>> entrySet = student.entrySet();
        for(Entry<String,Student> entry:entrySet){
            System.out.println("取得键: "+entry.getKey());
            System.out.println("对应的值为: "+entry.getValue().name);
        }
    }

    /*
     * 删除Map中的映射
     */
    public void testRemove(){
        Scanner sc = new Scanner(System.in);
        while(true){
            System.out.println("请输入要删除的学生ID: ");
            String stuID = sc.next();
            //判断输入的ID是否存在对应的学生对象
            Student stu = student.get(stuID);
            if(stu == null){
                System.out.println("输入的学生ID不存在! ");
                continue;
            }
            student.remove(stuID);
            System.out.println("成功删除学生"+stu.name);
            break;
        }
        testEntrySet();
    }

    /*
     * 使用put方法修改Map中已有的映射
     */
    public void testModify(){
        System.out.println("请输入要修改的学生ID: ");
        Scanner sc = new Scanner(System.in);
        while(true){
            String id = sc.next();
            Student stu = student.get(id);
            if(stu == null){
                System.out.println("ID不存在! ");
                continue;
            }
            System.out.println("当前学生是: "+stu.name);
            System.out.println("请输入新的学生: ");
            String name = sc.next();
            Student newStu = new Student(id,name);
            student.put(id, newStu);
```

```
                System.out.println("修改成功! ");
                break;
            }
        }

        /*
         * 测试Map中是否存在某个key值或value值
         */
        public void testContainsKey(){
            System.out.println("请输入学生ID: ");
            Scanner sc = new Scanner(System.in);
            String stuID = sc.next();
            //用containsKey()方法来判断是否存在某个key值
             System.out.println("输入的ID为: "+stuID+",在学生列表中是否存在: "+student.containsKey
(stuID));
            if(student.containsKey(stuID)){
                System.out.println("学生的姓名为: "+student.get(stuID).name);
            }

            System.out.println("请输入学生姓名: ");
            String name = sc.next();
            //用containsValue()方法来判断是否存在某个value值
            if(student.containsValue(new Student(null,name))){
                System.out.println("存在学生"+name);
            }else{
                System.out.println("学生不存在! ");
            }
        }

        public static void main(String[] args) {
            TestMap mt = new TestMap();
            mt.testPut();
            mt.testKeySet();
        }
    }
```

运行结果如图 2-8 所示。

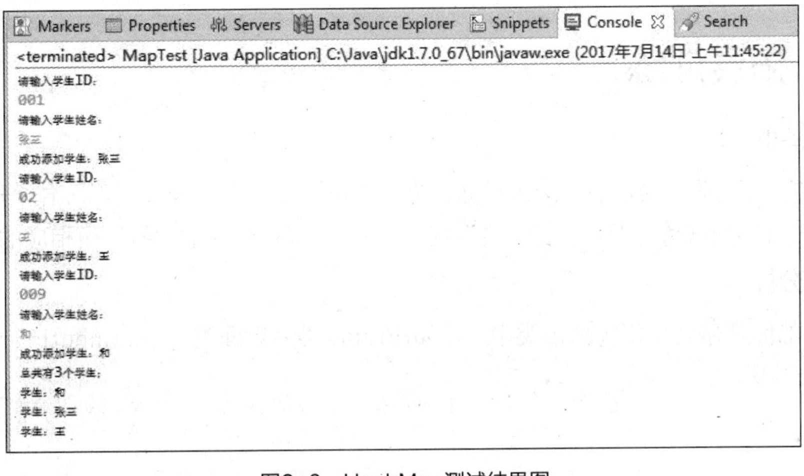

```
Markers  Properties  Servers  Data Source Explorer  Snippets  Console  Search
<terminated> MapTest [Java Application] C:\Java\jdk1.7.0_67\bin\javaw.exe (2017年7月14日 上午11:45:22)
请输入学生ID:
001
请输入学生姓名:
张三
成功添加学生: 张三
请输入学生ID:
02
请输入学生姓名:
王
成功添加学生: 王
请输入学生ID:
009
请输入学生姓名:
和
成功添加学生: 和
总共有3个学生:
学生: 和
学生: 张三
学生: 王
```

图2-8　HashMap测试结果图

2.4 拓展知识

上面主要对 Java 集合框架做了详细的介绍，包括 Collection 和 Map 两个接口及它们的抽象类和常用的具体实现类。下面主要介绍一下其他几个特殊的集合类，Vector、Stack、HashTable、ConcurrentHashMap 以及 CopyOnWriteArrayList。

1. Vector

前面我们已经提到，Java 设计者们在对之前的容器类进行重新设计时保留了一些数据结构，其中就有 Vector。用法上，Vector 与 ArrayList 基本一致，不同之处在于 Vector 使用了关键字 synchronized，将访问和修改向量的方法都变成同步的了，所以对于不需要同步的应用程序来说，类 ArrayList 比类 Vector 更高效。

2. Stack

Stack，栈类，是 Java2 之前引入的，继承自类 Vector。

3. HashTable

HashTable 和前面介绍的 HashMap 很类似，它也是一个散列表，存储的内容是键值对映射，不同之处在于，HashTable 是继承自 Dictionary 的，HashTable 中的函数都是同步的，这意味着它也是线程安全的，另外，HashTable 中 key 和 value 都不可以为 null。

上面的 3 个集合类都是在 Java2 之前推出的容器类，可以看到，尽管在使用中效率比较低，但是它们都是线程安全的。下面介绍两个特殊的集合类。

4. ConcurrentHashMap

Concurrent，并发，从名字就可以看出来 ConcurrentHashMap 是 HashMap 的线程安全版。同 HashMap 相比，ConcurrentHashMap 不仅保证了访问的线程安全性，而且在效率上与 HashTable 相比，也有较大的提高。

5. CopyOnWriteArrayList

CopyOnWriteArrayList 是一个线程安全的 List 接口的实现，它使用了 ReentrantLock 锁来保证在并发情况下提供高性能的并发读取。

2.5 拓展训练

1. 任务需求

实现省市两级联动。构建信息录入界面，完成人员基本信息的录入工作。其中籍贯中涉及的“省、市 / 县”能实现联动，即：选择“省”时，“市”会根据选择的“省”做相应的变换。

2. 任务分析

省市 / 县的信息存放在 CityMap 类中，InputFrame 类为界面类，TestInputFrame 为测试类。其类图如图 2-9 所示。

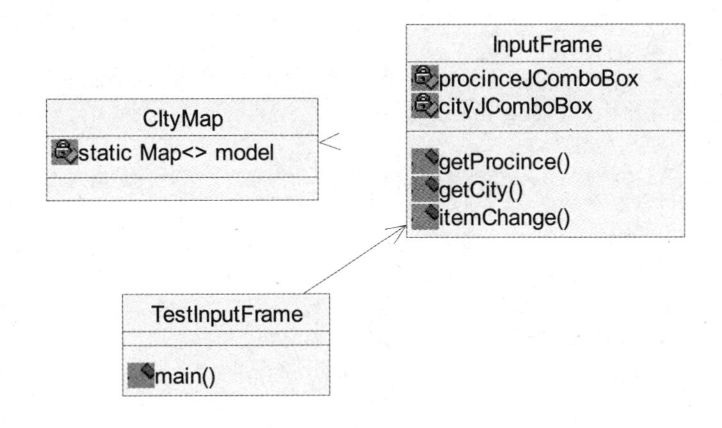

图2-9 类图

3. 任务实现

CityMap.java

```
package com.daiinfo.seniorjava.ken2.prolongation;
import java.util.LinkedHashMap;
import java.util.Map;
public class CityMap {
    /**
     * 全国（省、直辖市、自治区）映射集合
     */
    public static Map<String,String[]> model=new LinkedHashMap<String, String[]>();
    static{
        model.put("北京", new String[]{"北京"});
        model.put("上海", new String[]{"上海"});
        model.put("天津", new String[]{"天津"});
        model.put("重庆", new String[]{"重庆"});
        model.put("黑龙江", new String[]{"哈尔滨","齐齐哈尔","牡丹江","大庆","伊春","双鸭
山","鹤岗","鸡西","佳木斯","七台河","黑河","绥化","大兴安岭"});
        model.put("吉林", new String[]{"长春","延边","吉林","白山","白城","四平","松原","
辽源","大安","通化"});
        model.put("辽宁", new String[]{"沈阳","大连","葫芦岛","旅顺","本溪","抚顺","铁岭
","辽阳","营口","阜新","朝阳","锦州","丹东","鞍山"});
        model.put("内蒙古", new String[]{"呼和浩特","呼伦贝尔","锡林浩特","包头","赤峰","
海拉尔","乌海","鄂尔多斯","通辽"});
        model.put("河北", new String[]{"石家庄","唐山","张家口","廊坊","邢台","邯郸","沧州
","衡水","承德","保定","秦皇岛"});
        model.put("河南", new String[]{"郑州","开封","洛阳","平顶山","焦作","鹤壁","新乡
","安阳","濮阳","许昌","漯河","三门峡","南阳","商丘","信阳","周口","驻马店"});
        model.put("山东", new String[]{"济南","青岛","淄博","威海","曲阜","临沂","烟台","
枣庄","聊城","济宁","菏泽","泰安","日照","东营","德州","滨州","莱芜","潍坊"});
        model.put("山西", new String[]{"太原","阳泉","晋城","晋中","临汾","运城","长治","
朔州","忻州","大同","吕梁"});
        model.put("江苏", new String[]{"南京","苏州","昆山","南通","太仓","吴县","徐州","
宜兴","镇江","淮安","常熟","盐城","泰州","无锡","连云港","扬州","常州","宿迁"});
```

```
            model.put("安徽", new String[]{"合肥","巢湖","蚌埠","安庆","六安","滁州","马鞍山","阜阳","宣城","铜陵","淮北","芜湖","亳州","宿州","淮南","池州"});
            model.put("陕西", new String[]{"西安","韩城","安康","汉中","宝鸡","咸阳","榆林","渭南","商洛","铜川","延安"});
            model.put("宁夏", new String[]{"银川","固原","中卫","石嘴山","吴忠"});
            model.put("甘肃", new String[]{"兰州","白银","庆阳","酒泉","天水","武威","张掖","甘南","临夏","平凉","定西","金昌"});
            model.put("青海", new String[]{"西宁","海北","海西","黄南","果洛","玉树","海东","海南"});
            model.put("湖北", new String[]{"武汉","宜昌","黄冈","恩施","荆州","神农架","十堰","咸宁","襄樊","孝感","随州","黄石","荆门","鄂州"});
            model.put("湖南", new String[]{"长沙","邵阳","常德","郴州","吉首","株洲","娄底","湘潭","益阳","永州","岳阳","衡阳","怀化","韶山","张家界"});
            model.put("浙江", new String[]{"杭州","湖州","金华","宁波","丽水","绍兴","雁荡山","衢州","嘉兴","台州","舟山","温州"});
            model.put("江西", new String[]{"南昌","萍乡","九江","上饶","抚州","吉安","鹰潭","宜春","新余","景德镇","赣州"});
            model.put("福建", new String[]{"福州","厦门","龙岩","南平","宁德","莆田","泉州","三明","漳州"});
            model.put("贵州", new String[]{"贵阳","安顺","赤水","遵义","铜仁","六盘水","毕节","凯里","都匀"});
            model.put("四川", new String[]{"成都","泸州","内江","凉山","阿坝","巴中","广元","乐山","绵阳","德阳","攀枝花","雅安","宜宾","自贡","甘孜州","达州","资阳","广安","遂宁","眉山","南充"});
            model.put("广东", new String[]{"广州","深圳","潮州","韶关","湛江","惠州","清远","东莞","江门","茂名","肇庆","汕尾","河源","揭阳","梅州","中山","德庆","阳江","云浮","珠海","汕头","佛山"});
            model.put("广西", new String[]{"南宁","桂林","阳朔","柳州","梧州","玉林","桂平","贺州","钦州","贵港","防城港","百色","北海","河池","来宾","崇左"});
            model.put("云南", new String[]{"昆明","保山","楚雄","德宏","红河","临沧","怒江","曲靖","思茅","文山","玉溪","昭通","丽江","大理"});
            model.put("海南", new String[]{"海口","三亚","儋州","琼山","通什","文昌"});
            model.put("新疆", new String[]{"乌鲁木齐","阿勒泰","阿克苏","昌吉","哈密","和田","喀什","克拉玛依","石河子","塔城","库尔勒","吐鲁番","伊宁"});
        }
    }
```

InputFrame.java

```
package com.daiinfo.seniorjava.ken2.prolongation;
import java.awt.BorderLayout;
import java.awt.FlowLayout;
import java.awt.GridLayout;
import java.awt.Panel;
import java.awt.event.ActionEvent;
import java.awt.event.ActionListener;
import java.awt.event.ItemEvent;
import java.awt.event.ItemListener;
import java.util.Map;
import java.util.Set;
import javax.swing.DefaultComboBoxModel;
import javax.swing.JButton;
import javax.swing.JComboBox;
import javax.swing.JFrame;
```

```
import javax.swing.JLabel;
import javax.swing.JPanel;
import javax.swing.JScrollPane;
import javax.swing.JTextField;
import javax.swing.JTextArea;
import javax.swing.border.TitledBorder;
public class InputFrame extends JFrame implements ActionListener{

    final long serialVersionUID = -4595347311922711984L;
    private JTextField nameJTextField;
    private JTextField addressJTextField;
    private JComboBox sexJComboBox;
    private JTextField emailJTextField;
    private JComboBox provinceJComboBox;
    private JComboBox cityJComboBox;
    private JLabel  nameJLabel;
    private JLabel  sexJLabel;
    private JLabel  provinceJLabel;
    private JLabel  cityJLabel;
    private JLabel  emailJLabel;
    private JLabel  addressJLabel;
    private JLabel  nativePlaceJLabel;
    private JButton saveButton;
    private JButton cancelButton;
    private JTextArea infoArea;


    /**
     * 构造函数
     */
    public InputFrame(String title) {
        super(title);
        getContentPane().setLayout(null);
        setBounds(100, 100, 500, 660);
        setDefaultCloseOperation(JFrame.EXIT_ON_CLOSE);

        nameJTextField=new JTextField(20);
        addressJTextField=new JTextField();
        emailJTextField=new JTextField(30);

        sexJComboBox=new JComboBox();
        sexJComboBox.setModel(new javax.swing.DefaultComboBoxModel(new String[] {"男","女"}));

        provinceJComboBox=new JComboBox<>();
        cityJComboBox=new JComboBox<>();

        nameJLabel=new JLabel("姓名");
        sexJLabel=new JLabel("性别");
        nativePlaceJLabel=new JLabel("籍贯");
        provinceJLabel=new JLabel("省");
```

```
            cityJLabel=new JLabel("市");
            addressJLabel=new JLabel("通讯地址");
            emailJLabel=new JLabel("邮箱");

            final JPanel pan1 = new JPanel();
            pan1.setLayout(new GridLayout());
            pan1.setBorder(new TitledBorder(null, "基本信息", TitledBorder.DEFAULT_JUSTIFICATION,
TitledBorder.DEFAULT_POSITION, null, null));
            pan1.setBounds(12, 12, 418, 70);
            getContentPane().add(pan1);
            pan1.add(nameJLabel);
            pan1.add(nameJTextField);
            pan1.add(sexJLabel);
            pan1.add(sexJComboBox);

            final JPanel pan2 = new JPanel();
            pan2.setLayout(new GridLayout());
            pan2.setBounds(12, 98, 418, 70);
            pan2.setBorder(new TitledBorder(null, "籍贯", TitledBorder.DEFAULT_JUSTIFICATION,
TitledBorder.DEFAULT_POSITION, null, null));
            getContentPane().add(pan2);
            pan2.add(provinceJLabel);
            pan2.add(provinceJComboBox);
            pan2.add(cityJLabel);
            pan2.add(cityJComboBox);


            provinceJComboBox.addItemListener(new ItemListener(){
                public void itemStateChanged(final ItemEvent e) {      // 选项状态更改事件
                    itemChange();
                }
            });

            provinceJComboBox.setModel(new DefaultComboBoxModel(getProvince())); // 添加
省份信息
            String province=(String)getProvince()[0];
            cityJComboBox.setModel(new DefaultComboBoxModel(getCity(province)));

            final JPanel pan3 = new JPanel();
            pan3.setLayout(new GridLayout());
            pan3.setBounds(12, 178, 418, 70);
            pan3.setBorder(new TitledBorder(null, "通讯地址", TitledBorder.DEFAULT_JUSTIFICATION,
TitledBorder.DEFAULT_POSITION, null, null));
            getContentPane().add(pan3);
            pan3.add(addressJLabel);
            pan3.add(addressJTextField);
            pan3.add(emailJLabel);
            pan3.add(emailJTextField);

            final JPanel pan4 = new JPanel();
```

```
        pan4.setLayout(null);
        pan4.setBounds(12, 278, 418, 70);
        pan4.setBorder(new TitledBorder(null, "", TitledBorder.DEFAULT_JUSTIFICATION,
TitledBorder.DEFAULT_POSITION, null, null));
        getContentPane().add(pan4);
        saveButton=new JButton("保存");
        saveButton.setBounds(80,20,60, 40);

        cancelButton=new JButton("取消");
        cancelButton.setBounds(280,20,60, 40);
        pan4.add(saveButton);
        pan4.add(cancelButton);
        saveButton.addActionListener(this);
        cancelButton.addActionListener(this);

        infoArea=new JTextArea("基本信息",100,200);
        infoArea.setBounds(12, 360, 420, 200);
        add(infoArea);


    }


    /**
     * 获取省、直辖市、自治区
     *
     * @return
     */
    public Object[] getProvince() {
        Map<String, String[]> map = CityMap.model;// 获取省份信息保存到Map中
        Set<String> set = map.keySet(); // 获取Map集合中的键，并以Set集合返回
        Object[] province = set.toArray(); // 转换为数组
        return province; // 返回获取的省份信息
    }

    /**
     * 获取指定省对应的市/县
     *
     * @param selectProvince
     * @return
     */
    public String[] getCity(String selectProvince) {
        Map<String, String[]> map = CityMap.model; // 获取省份信息保存到Map中
        String[] arrCity = map.get(selectProvince); // 获取指定键的值
        return arrCity; // 返回获取的市/县
    }
    private void itemChange() {
        String selectProvince = (String) provinceJComboBox.getSelectedItem();
        cityJComboBox.removeAllItems(); // 清空市/县列表
        String[] arrCity = getCity(selectProvince); // 获取市/县
```

```
            cityJComboBox.setModel(new DefaultComboBoxModel(arrCity)); //重新添加市/县列表的值
        }

        @Override
        public void actionPerformed(ActionEvent e) {
            // TODO Auto-generated method stub
            if(e.getSource()==saveButton){
                infoArea.append("\n");
                String string="";
                string+="\n"+"姓名: "+nameJTextField.getText();
                string+="\n"+"性别: "+sexJComboBox.getSelectedItem();
                string+="\n"+"籍贯: "+provinceJComboBox.getSelectedItem()+"省"+cityJComboBox.
getSelectedItem()+"市/县";
                string+="\n"+"通讯地址: "+addressJTextField.getText();
                string+="\n"+"邮箱: "+emailJTextField.getText();
                infoArea.append(string);
            }

            if(e.getSource()==cancelButton){
                infoArea.setText("基本信息");
            }
        }
    }
```

TestInputFrame.java

```
    package com.daiinfo.seniorjava.ken2.prolongation;
    import java.awt.EventQueue;
    import javax.swing.UIManager;
    public class TestInputFrame {
        public static void main(String[] args) {
            // TODO Auto-generated method stub
            EventQueue.invokeLater(new Runnable() {
                public void run() {
                    try {
                        //UIManager.setLookAndFeel("com.sun.Java.swing.plaf.nimbus.
NimbusLookAndFeel");
                        InputFrame frame = new InputFrame("信息录入界面");
                        frame.setVisible(true);
                    } catch (Exception e) {
                        e.printStackTrace();
                    }
                }
            });
        }
    }
```

运行结果如图 2-10 所示。

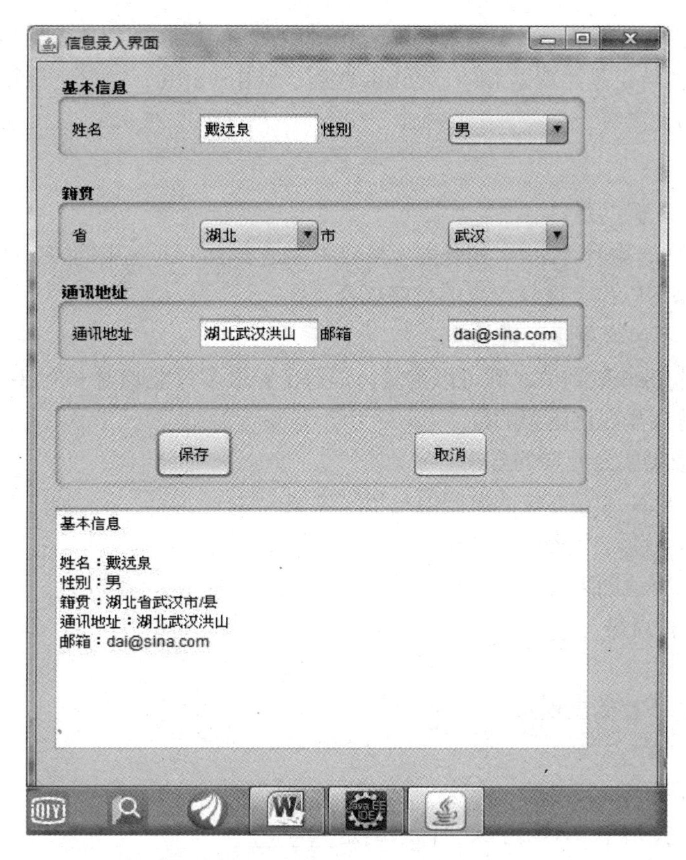

图2-10　省市两级联动运行结果图

2.6　课后小结

到这里，对于 Java 集合框架的总结就结束了，还有很多集合类没有在这里提到，更多的还是需要大家自己去查去用。通过阅读源码，查阅资料，收获会很大。

- Java 集合框架主要包括 Collection 和 Map 两种类型。其中 Collection 又有 3 种子类型，分别是 List、Set、Queue。Map 中存储的主要是键值对映射。
- 规则集 Set 中存储的是不重复的元素，线性表中可以存储重复的元素，Queue 队列描述的是先进先出的数据结构，可以用 LinkedList 来实现队列。
- 效率上，规则集比线性表更高效。
- ArrayList 主要是用数组来存储元素，LinkedList 主要是用链表来存储元素，HashMap 的底层实现主要是借助数组 + 链表 + 红黑树来实现。
- Vector、HashTable 等集合类效率比较低，但都是线程安全的。包 Java.util.concurrent 下包含了大量线程安全的集合类，效率上有较大提升。

2.7　课后习题

一、填空题

1. List 接口的特点是元素________（有 | 无）顺序，________（可以 | 不可以）重复；

2. Set 接口的特点是元素________（有 | 无）顺序，________（可以 | 不可以）重复；

3. Map 接口的特点是元素是 key、value 映射，其中 value________重复，key________重复（可以 | 不可以）。

二、选择题

1. 下面说法不正确的是________。

A. 列表（List）、集合（Set）和映射（Map）都是 Java.util 包中的接口

B. List 接口是可以包含重复元素的有序集合

C. Set 接口是不包含重复元素的集合

D. Map 接口将键映射到值，键可以重复，但每个键最多只能映射一个值

2. 下面关于迭代器说法错误的是________。

A. 迭代器是取出集合元素的方式

B. 迭代器的 hasNext() 方法的返回值是布尔类型

C. List 集合有特有迭代器

D. next() 方法将返回集合中的上一个元素

3. Set 集合的特点是________。

A. 元素有序

B. 元素无序，不存储重复元素

C. 存储重复元素

D. Set 集合都是线程安全的

4. 下面类或者接口中，不属于集合体系的是________。

A. Java.util.Collections　　B.Java.util.Map

C. Java.util.Vector　　D. Java.util.Hashtable

5. 以下能以键值对的方式存储对象的接口是________。

A. Java.util.Collection　　B. Java.util.Map

C. Java.util.HashMap　　D. Java.util.Set

6. 在 Java 中，________类可用于创建链表数据结构的对象。

A. LinkedList　　B. ArrayList

C. Collection　　D. HashMap

7. 代码的功能为：对于一个存放 Person 对象的 ArrayList 进行循环遍历，并输出每个 Person 对象的 idCard 和 userName。

```
public class Person{
     private Long idCard;
     pirvate String userName;
  //以下是getter和setter方法    //省略
  }
List list=new ArrayList();
Person p1=new Person();
p1.setIdCard(new Long(1001));
p1.setUserName("terry");
Person p2=new Person();
p2.setIdCard(new Long(1002));
p2.setUserName("tom");
```

```
list.add(p1);
 list.add(p2);
for( 位置① ){
System.out.println(person.getIdCard()+":"+person.getUserName());
 }
```

那么位置①处的代码为________。

A. List list:person　　B. List list:Person

C. Person person:List　　D. Person person:list

三、简答题

1. Java 集合框架是什么？说出一些集合框架的优点。
2. Java 集合框架的基础接口有哪些？
3. Iterator 是什么？
4. 遍历一个 List 有哪些不同的方式？
5. 如何决定选用 HashMap 还是 TreeMap ？
6. ArrayList 和 LinkedList 有何区别？
7. ArrayList 和 Vector 有何异同点？

2.8 上机实训

实训一　Map 集合的各种使用方法附带 keySet() 和 entrySet() 的使用。

设有一学生类 Student.Java，将学生信息存入 Map 中。使用 keySet() 和 entrySet() 实现对学生信息的“增删改查”操作。

实训二　已知某学校的教学课程内容安排如下。

老师	课程
张和	Java 语言
李纪	C 语言
王桔	JSP
胡海	Oracle 数据库
刘山	网页设计
郭富	JSP
付格	Linux

完成下列要求。

（1）使用一个 Map，以老师的名字作为键，以课程名作为值，表示上述课程安排。

（2）增加了一位新老师 Allen 教 JDBC。

（3）Lucy 改为教 CoreJava　put 方法 。

（4）遍历 Map，输出所有的老师及老师教授的课程（Set<Map.Entry<String,String>>、Set<String> get(key)）。

（5）利用 Map，输出所有教 JSP 的老师。

PART03

知识领域3

Java反射机制

知识目标

1. 理解Java的反射机制。
2. 掌握Java反射机制的应用。

■ 能力目标

熟练使用Java反射机制的API编写应用程序。

■ 素质目标

1. 能够阅读科技文档和撰写分析文档。
2. 能够查阅JDK API。

3.1 应用场景

在一些开源框架里，如 Spring、Struts、Hibernate、MyBatis 等，应用程序会提供一个配置文件，如 xml 文件或者 properties，然后在 Java 类里面解析 xml 或 properties 里面的内容，得到一个字符串，然后用反射机制，根据这个字符串获得某个类的 Class 实例，这样就可以动态配置一些东西，不用每一次都在代码里面去 new 或者做其他的事情，以后要改的话直接改配置文件，代码维护起来就很方便了。同时有时候要适应某些需求，Java 类里面不一定能直接调用另外的方法，这时候也可以通过反射机制来实现。

应用程序通过读取配置文件来获取到指定名称的类的字节码文件并加载其中的内容进行调用，对一个类文件进行解析，就可以取得任意一个已知名称的 class 的内部信息，包括该类的修饰（如 public、static 等）、其继承的父类（如 Object）、其实现的接口（如 Serializable），也包括其属性和方法的所有信息，并可于运行时改变属性内容或调用方法。

3.2 相关知识

3.2.1 Java 反射机制的概念

在 Java 运行状态中，对于任意一个类，我们都能够知道这个类的所有属性和方法，对于任意一个对象，我们都能够调用它的任意一个方法。这种动态获取信息以及动态调用对象方法的功能称为 Java 语言的反射机制。

3.2.2 反射机制的功能

反射机制的功能如下。

- 在运行时判定任意一个对象所属的类；
- 在运行时构造任意一个类的对象；
- 在运行时判定任意一个类所具有的成员变量和方法；
- 在运行时调用任意一个对象的方法；
- 生成动态代理。

3.2.3 Java 反射机制的相关 API

Java.lang.reflect 包提供了用于获取类和对象的反射信息的类和接口。反射 API 允许对程序访问有关加载类的字段、方法和构造函数的信息进行编程访问。它允许在安全限制内使用反射的字段、方法和构造函数对其底层对等进行操作。

- Java.lang.Class;　　　　　　　// 类
- Java.lang.reflect.Constructor;　// 构造方法
- Java.lang.reflect.Field;　　　　// 类的成员变量
- Java.lang.reflect.Method;　　　// 类的方法
- Java.lang.reflect.Modifier;　　 // 访问权限

Class 类提供了大量的实例方法来获取该 Class 对象所对应的详细信息，Class 类大致包含如下方法，其中每个方法都包含多个重载版本，因此我们只是做简单的介绍，详细请参考 JDK 文档，其中一般常见主要方法如表 3-1 所示。

表3-1 Class类主要方法

序号	获取内容	方法签名
1	构造器	Constructor<T> getConstructor(Class<?>…parameterTypes)
2	包含的方法	Method getMethod(String name，Class<?>…parameterTypes)
3	包含的属性	Field getField(String name)
4	包含的 Annotation	<A extends Annotation> A getAnnotation (Class<A> annotationClass)
5	内部类	Class<?>[ ] getDeclaredClasses( )
6	外部类	Class<?> getDeclaringClass( )
7	所实现的接口	Class<?>[ ] getInterfaces( )
8	修饰符	int getModifiers( )
9	所在包	Package getPackage( )
10	类名	String getName( )
11	简称	String getSimpleName( )

3.2.4 使用反射机制的步骤

导入 Java.lang.relfect 包需遵循以下 3 个步骤。

- 第一步是获得你想操作的类的 Java.lang.Class 对象；
- 第二步是调用诸如 getDeclaredMethods 的方法；
- 第三步是使用反射 API 来操作这些信息。

3.2.5 反射机制的应用场景

Java 反射机制是在运行状态中，对于任意一个类，都能够知道这个类的所有属性和方法；对于任意一个对象，都能够调用它的任意一个方法。常见的应用如下。

- 逆向代码，例如反编译；
- 与注解相结合的框架，例如 Retrofit；
- 单纯的反射机制应用框架，例如 EventBus 2.x；
- 动态生成类框架，例如 Gson。

3.2.6 反射机制的优缺点

反射机制可以实现动态创建对象和编译，体现出很大的灵活性（特别是在 J2EE 的开发中它的灵活性就表现得十分明显）。通过反射机制我们可以获得类的各种内容，进行了反编译。对于 Java 这种先编译再运行的语言来说，反射机制可以使代码更加灵活，更加容易实现面向对象。总结如下。

- 优点：运行期类型的判断，动态类加载，动态代理使用反射。
- 缺点：性能是一个问题，反射相当于一系列解释操作，通知 jvm 要做的事情，性能比直接的 Java 代码要慢很多。

3.3 任务实施

任务　使用反射机制获取类的相关信息

1. 任务需求

设有一个类 Person，使用反射机制获取该类的相关信息。

2. 任务分析

实体类 Person 存放个人的基本信息，测试类 TestPerson 使用反射机制获取 Person 类或其对象的封装信息。其类图如图 3-1 所示。

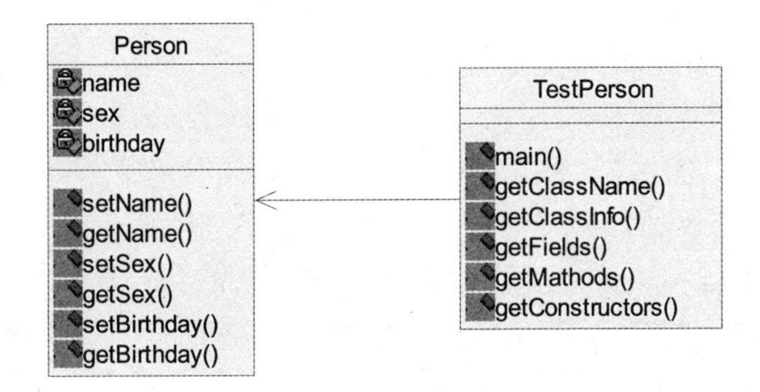

图3-1　类图

3. 任务实现

Person.java

```
package com.daiinfo.seniorjava.ken3.implement;
public class Person {
    public String name;
    public String sex;
    String birthday;

    /**
     * 构造函数
     */
    public Person(){

    }

    /**
     * 构造函数
     * @param name
     */
    public Person(String name){
        this.name=name;
    }

    /**
```

```
     * 构造函数
     * @param name
     * @param sex
     * @param birthday
     */
    public Person(String name,String sex,String  birthday) {
        this.name=name;
        this.sex=sex;
        this.birthday=birthday;
    }


    public String getName() {
        return name;
    }
    public void setName(String name) {
        this.name = name;
    }
    public String getSex() {
        return sex;
    }
    public void setSex(String sex) {
        this.sex = sex;
    }
    public String getBirthday() {
        return birthday;
    }
    public void setBirthday(String birthday) {
        this.birthday = birthday;
    }
}
```

（1）class 对象的获取

```
public static void getClassName() {
        // 第一种方式：通过对象getClass方法
        Person person = new Person("张", "男", "2010-09-10");
        Class<?> class1 = person.getClass();
        // 第二种方式：通过类的class属性
        class1 = Person.class;
        try {
            // 第三种方式：通过Class类的静态方法——forName()来实现
            class1 = Class.forName("com.hbliti.example3.Person");
        } catch (ClassNotFoundException e) {
            e.printStackTrace();
        }
}
```

（2）获取 class 对象的摘要信息

```
static void getClassInfo() {
        Class<?> class1 = null;
        try {
```

```
            class1 = Class.forName("com.hbliti.example3.Person");
        } catch (ClassNotFoundException e) {
            // TODO Auto-generated catch block
            e.printStackTrace();
        }
        boolean isPrimitive = class1.isPrimitive();// 判断是否是基础类型
        boolean isArray = class1.isArray();// 判断是否是集合类
        boolean isAnnotation = class1.isAnnotation();// 判断是否是注解类
        boolean isInterface = class1.isInterface();// 判断是否是接口类
        boolean isEnum = class1.isEnum();// 判断是否是枚举类
        boolean isAnonymousClass = class1.isAnonymousClass();//判断是否是匿名内部类
        boolean isAnnotationPresent = class1
                .isAnnotationPresent(Deprecated.class);//判断是否被某个注解类修饰
        String className = class1.getName();// 获取class名字，包含包名路径
        Package aPackage = class1.getPackage();// 获取class的包信息
        String simpleName = class1.getSimpleName();// 获取class类名
        int modifiers = class1.getModifiers();// 获取class访问权限
        Class<?>[] declaredClasses = class1.getDeclaredClasses();// 内部类
        Class<?> declaringClass = class1.getDeclaringClass(); // 外部类
        System.out.print(aPackage + "\t" + simpleName + "\t" + modifiers);
}
```

（3）获取 class 对象的属性

```
static void getFields() {
        Class<?> class1 = null;
        try {
            class1 = Class.forName("com.hbliti.example3.Person");
        } catch (ClassNotFoundException e) {
            // TODO Auto-generated catch block
            e.printStackTrace();
        }
        Field[] allFields = class1.getDeclaredFields();// 获取class对象的所有属性
        Field[] publicFields = class1.getFields();// 获取class对象的public属性
        try {
            Field nameField = class1.getDeclaredField("name");// 获取class指定属性
            Field sexField = class1.getField("sex");// 获取class指定的public属性
        } catch (NoSuchFieldException e) {
            e.printStackTrace();
        }
        for (int i = 0; i < allFields.length; i++) {
            System.out.println(allFields[i]);
        }
    }
```

（4）获取 class 对象的方法

```
static void getMethods() {
        Class<?> class1 = null;
        try {
            class1 = Class.forName("com.hbliti.example3.Person");
        } catch (ClassNotFoundException e) {
            // TODO Auto-generated catch block
```

```
            e.printStackTrace();
        }
        Method[] methods = class1.getDeclaredMethods();// 获取class对象的所有声明方法
        Method[] allMethods = class1.getMethods();// 获取class对象的所有方法，包括父类的方法
        for (int i = 0; i < allMethods.length; i++) {
            System.out.println(allMethods[i]);
        }
}
```

（5）获取 class 对象的构造函数

```
    static void getConstructors(){
        Class<?> class1 = null;
        try {
            class1 = Class.forName("com.hbliti.example3.Person");
        } catch (ClassNotFoundException e) {
            // TODO Auto-generated catch block
            e.printStackTrace();
        }
        Constructor<?>[] allConstructors = class1.getDeclaredConstructors();//获取class
对象的所有声明构造函数
        Constructor<?>[] publicConstructors = class1.getConstructors();//获取class对象public构
造函数
        for (int i = 0; i < allConstructors.length; i++) {
            System.out.println(allConstructors[i]);
        }
    }
```

运行结果如图 3-2 所示。

```
Markers  Properties  Servers  Data Source Explorer  Snippets  Console ☒  Search
<terminated> ReflactTest [Java Application] C:\Java\jdk1.7.0_67\bin\javaw.exe (2017年6月26日 下午10:56:08)
public com.hbliti.example3.Person(java.lang.String,java.lang.String,java.lang.String)
package com.hbliti.example3     Person  1public java.lang.String com.hbliti.example3.Person.name
public java.lang.String com.hbliti.example3.Person.sex
java.lang.String com.hbliti.example3.Person.birthday
public java.lang.String com.hbliti.example3.Person.getName()
public void com.hbliti.example3.Person.setName(java.lang.String)
public void com.hbliti.example3.Person.setSex(java.lang.String)
public java.lang.String com.hbliti.example3.Person.getSex()
public void com.hbliti.example3.Person.setBirthday(java.lang.String)
public java.lang.String com.hbliti.example3.Person.getBirthday()
public final void java.lang.Object.wait(long,int) throws java.lang.InterruptedException
public final native void java.lang.Object.wait(long) throws java.lang.InterruptedException
public final void java.lang.Object.wait() throws java.lang.InterruptedException
public boolean java.lang.Object.equals(java.lang.Object)
public java.lang.String java.lang.Object.toString()
public native int java.lang.Object.hashCode()
public final native java.lang.Class java.lang.Object.getClass()
public final native void java.lang.Object.notify()
public final native void java.lang.Object.notifyAll()
```

图3-2 测试类运行结果图

3.4 拓展知识

Java 的反射机制是 Java 特性之一，反射机制是构建框架技术的基础所在。

在设计模式学习当中，学习抽象工厂的时候就用到了反射来更加方便地读取数据库链接字符

串等。Java 的配置文件为 .properties，称作属性文件。通过反射读取里边的内容，这样代码是固定的，但是配置文件的内容我们可以改，这样使我们的代码灵活了很多。

典型的除了 Hibernate 之外，还有 Spring 也用到很多反射机制。灵活掌握 Java 反射机制，对大家以后学习框架技术有很大的帮助。

3.5 拓展训练

1. 任务需求

练习 Java 反射机制在工厂模式中的应用，更好地理解工厂设计模式。

2. 任务分析

简单工厂模式（simple factory）设计包括接口 Car，类 Benz、Bmw、Bike。

具体类图如图 3-3 所示。

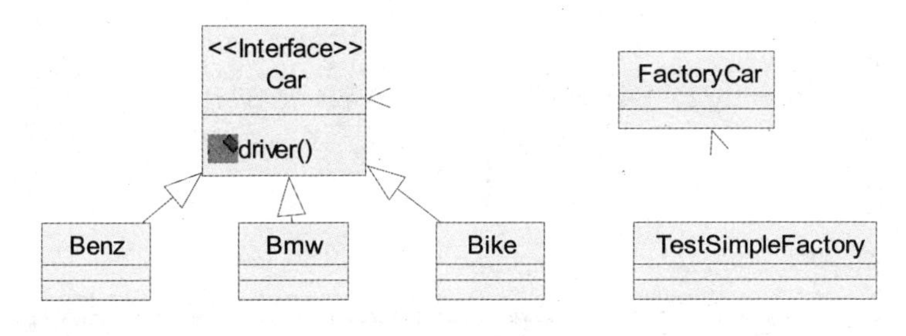

图3-3 类图

3. 任务实现

Car.java

```
package com.daiinfo.seniorjava.ken3.prolongation;
/**
 * 车的父类
 * @author daiyuanquan
 *
 */
interface Car {

    /**
     * 开车
     */
    public void driver();
}
```

Benz.java

```
package com.daiinfo.seniorjava.ken3.prolongation;
public class Benz implements Car {
    @Override
    public void driver() {
        System.out.println("今天咱开奔驰！");
```

```
        }
    }
```

Bmw.java

```
    package com.daiinfo.seniorjava.ken3.prolongation;
    public class Bmw implements Car {
        @Override
        public void driver() {
            System.out.println("今天开宝马吧！");
        }
    }
```

Bike.java

```
    package com.daiinfo.seniorjava.ken3.prolongation;
    public class Bike implements Car{
        @Override
        public void driver() {
            System.out.println("唉，现在经济危机，只能骑自行车了呀！");
        }
    }
```

FactoryCar.java

```
    package com.daiinfo.seniorjava.ken3.prolongation;
    public class FactoryCar {
        public static Car driverCar(String s) throws Exception {
            if (s.equalsIgnoreCase("Benz")) {// 判断传入参数，返回不同的实现类
                return new Benz();
            } else if (s.equalsIgnoreCase("Bmw")) {
                return new Bmw();
            } else if (s.equalsIgnoreCase("Bike")) {
                return new Bike();
            } else {
                throw new Exception();// 抛出异常
            }
        }
    }
```

TestSimpleFactory.java

```
    package com.daiinfo.seniorjava.ken3.prolongation;
    public class TestSimpleFactory {
        public static void main(String[] args) {// java程序主入口处
            try {

                System.out.println("经理，今天开什么车呀？");
                Car car = FactoryCar.driverCar("Bike");// 调用方法返回车的实例
                car.driver();// 调用方法开车
            } catch (Exception e) {// 捕获异常
                System.out.println("开车出现问题……");
            } finally {// 代码总被执行
```

```
            System.out.println("……");
        }
    }
}
```

运行结果如图 3-4 所示。

Markers　Properties　Servers　Data Source Explorer　Snippets　Console　Search

<terminated> TestSimpleFactory [Java Application] C:\Java\jdk1.7.0_67\bin\javaw.exe (2017年7月25日 下午12:40:01)

经理，今天开什么车呀？

唉，现在经济危机，只能骑自行车了呀！

……

图3-4　简单工厂模式运行结果图

上面代码中，定义了一个车（Car）接口，接着分别定义了奔驰（Benz）、宝马（Bmw）和自行车（Bike）实现类，实现了 driver() 方法。

如果我们现在不用工厂模式，在调用奔驰车方法时需要 new 一个奔驰（Benz）的对象，使用宝马车方法时，需要 new 一个宝马（Bmw）类的对象。这样在 new 对象时，就需要考虑到接口和子类的实现方式，增加了代码的耦合度。在代码中使用 new 操作其实也是一种硬编码方式。

使用简单工厂模式解决这个问题（上面的代码），封装一个工厂类，把 new 对象的操作放在这个工厂类中，这样我们在调用子类里面的方法时，并不需要考虑子类的实现方式，只需要调用工厂类，让工厂类给我们实现 new 对象的过程，使子类与外界操作没有联系，降低代码耦合性。

大家应该也发现了简单工厂模式有个很大的弊端，就是现在要实现很多车的 driver() 方法，那么工厂类里面就要不断地添加 new 对象的操作，修改里面的代码。其类图如图 3-5 所示。

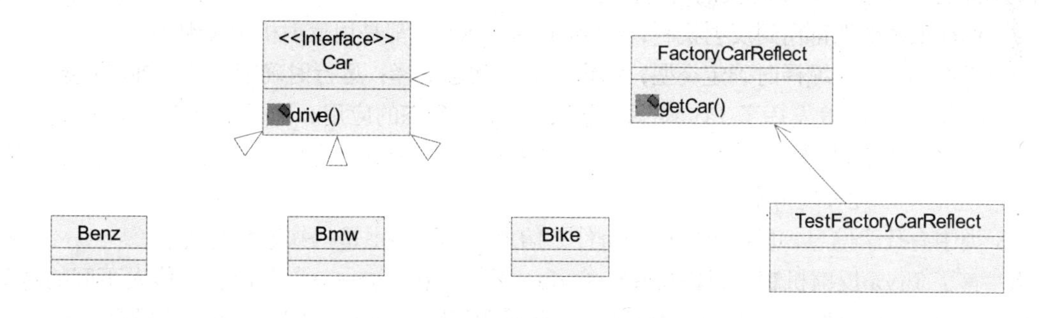

图3-5　类图

现在采用反射机制来修改工厂方法。

FactoryCarReflect.java

```
package com.daiinfo.seniorjava.ken3.prolongation;
public class FactoryCarReflect {
    public Object getCar(String className) throws Exception{
        Class<?> cls = Class.forName(className);
        Object obj = cls.newInstance();
        return obj;
    }
}
```

TestFactoryCarReflect.java

```
package com.daiinfo.seniorjava.ken3.prolongation;
public class TestFactoryCarReflect {
    public static void main(String[] args) throws Exception{
        FactoryCarReflect factory = new FactoryCarReflect();
        Car car = (Car)factory.getCar("com.daiinfo.seniorjava.ken3.prolongation.Benz");
        car.driver();
    }
}
```

运行结果如图 3-6 所示。

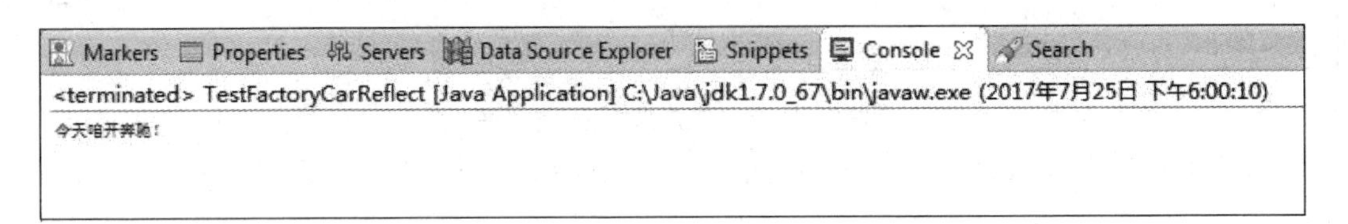

图3-6　采用反射机制的工厂模式的运行结果图

采用反射后，无论添加多少个子类，工厂类中的代码都不需要修改，只需要在操作的时候传入子类的类路径（包 . 类）就可以了，实现了各个业务逻辑之间的完全分离，代码耦合性进一步降低，很好地解决了上面的问题。

3.6　课后小结

1. 反射机制指的是程序在运行时能够获取自身的信息。在 Java 中，只要给定类的名字，那么就可以通过反射机制来获得类的所有信息。

2. 现在很多框架都用到反射机制，hibernate、struts 都是用反射机制实现的。

3. 静态编译：在编译时确定类型，绑定对象；动态编译：运行时确定类型，绑定对象。

动态编译最大限度发挥了 Java 的灵活性，体现了多态的应用，降低了类之间的藕合性。一句话，反射机制的优点就是可以实现动态创建对象和编译，体现出很大的灵活性，特别是在 J2EE 的开发中它的灵活性表现得十分明显。

4. 类中有什么信息，反射就可以获得什么信息，不过前提是得知道类的名字。

5. 有了 Java 反射机制，只需要写一个 dao 类，4 个方法——增删改查，传入不同的对象，无须为每一个表都创建 dao 类，反射机制会自动帮我们完成剩下的事情，反射机制就是专门帮我们做那些重复的有规则的事情。

3.7　课后习题

一、填空题

1. 反射中常常用的几个类是________、________、________、________和________。
2. Field fields[ ] = classType.getDeclaredFields(); 的作用是________。
3. Java 可以在运行时动态获取某个类的类信息，这就是________。
4. Class clazz = Class.forName("com.yano.reflect.Person"); 的作用是________。

二、选择题

1. 下列哪个选项不属于 Java 程序加载类的步骤________。

A. 加载：读取 class 文件

B. 连接：验证内部结构，为静态资源分配空间，处理非静态引用

C. 初始化：将代码放到代码区，初始化静态成员，将静态和非静态分离

D. 创建对象：为该类创建一个普通的对象

2. 使用反射机制获取一个类的属性，下列关于 getField() 方法说法正确的是________。

A. 该方法需要一个 String 类型的参数来指定要获取的属性名

B. 该方法只能获取私有属性

C. 该方法只能获取公有属性

D. 该方法可以获取私有属性，但使用前必须先调用 setAccessible(true)

3. 下列关于通过反射方式获取方法并执行的过程说法正确的是________。

A. 通过对象名 . 方法名 (参数列表) 的方式调用该方法

B. 通过 Class.getMethod（方法名，参数类型列表）的方式获取该方法

C. 通过 Class.getDeclaredMethod（方法名，参数类型列表）获取私有方法

D. 通过 invoke(对象名，参数列表) 方法来执行一个方法

4. 关于反射机制下列说法错误的是________。

A. 反射机制指的是在程序运行过程中，通过 .class 文件加载并使用一个类的过程

B. 反射机制指的是在程序编译期间，通过 .class 文件加载并使用一个类的过程

C. 反射可以获取类中所有的属性和方法

D. 暴力反射可以获取类中私有的属性和方法

三、简答题

1. 描述反射机制的作用。举几个反射的应用实例。

2. 简述 Java 反射中 API 的主要类及作用。

3. 简述 Java 反射的步骤。

3.8 上机实训

本例实现从配置文件中读取类，并显示其构造函数，配置文件代码如下：

config.properties

```
classname=com.daiinfo.seniorjava.ken3.training.Point
method=move
args=12,24,"武汉"
argsType=int,int,String
```

Point.java

```
package com.daiinfo.seniorjava.ken3.training;
public class Point {
    public int x;
    private int y;
    private String name;//该点的名称
    /**
```

```
 * 无参构造函数
 */
public Point(){

}

/**
 * 一个参数的构造函数
 * @param name
 */
public Point(String name){
    this.name=name;
}

/**
 * 1个参数的构造函数
 * @param x
 */
public Point(int x){
    this.x=x;
}

/**
 * 2个参数的故障函数
 * @param x
 * @param y
 */
public  Point(int x ,int y){
    this.x=x;
    this.y=y;
}

/**
 * 3个参数的构造函数
 * @param x
 * @param y
 * @param name
 */
public Point(int x,int y,String name){
    this.x=x;
    this.y=y;
    this.name=name;
}

/**
 * 移动点到新的位置，水平偏移量dx，垂直偏移量dy，并命名新的名称
 * @param dx
 * @param dy
 * @param nameString
 */
public void move(int dx,int dy,String nameString){
```

```
        x=x+dx;
        y=y+dy;
        name=nameString;
    }

    public int hashCode() {
        final int prime = 31;
        int result = 1;
        result = prime * result + x;
        result = prime * result + y;
        return result;
    }
    public int getX() {
        return x;
    }
    public void setX(int x) {
        this.x = x;
    }
    public int getY() {
        return y;
    }
    public void setY(int y) {
        this.y = y;
    }
    public boolean equals(Object obj) {
        if (this == obj)
            return true;
        if (obj == null)
            return false;
        if (getClass() != obj.getClass())
            return false;
        Point other = (Point) obj;
        if (x != other.x)
            return false;
        if (y != other.y)
            return false;
        return true;
    }
}
```

TestReflect.java

```
public class TestReflect {
    public static void main(String[] args) throws Exception {
        //TODO Auto-generated method stub

    }
}
```

PART04

知识领域4

Java泛型机制

知识目标

1. 理解Java泛型的应用场景、泛型类和泛型方法。
2. 掌握Java泛型类、泛型接口和泛型方法的使用。

■ 能力目标

熟练使用Java泛型类、泛型接口和泛型方法等编写相应的应用程序。

■ 素质目标

1. 能够阅读科技文档和撰写分析文档。
2. 能够查阅JDK API。
3. 增强学生团队协作能力。

4.1 应用场景

假定我们有这样一个需求：写一个排序方法，能够对整型数组、字符串数组甚至其他任何类型的数组进行排序，该如何实现呢？答案是可以使用 Java 泛型。

使用 Java 泛型的概念，我们可以写一个泛型方法来对一个对象数组排序。然后，调用该泛型方法来对整型数组、浮点数数组、字符串数组等进行排序。

在使用框架 SSH（Struts+Spring+Hibernate）开发一个应用系统中，常使用 DAO（Date Access Object）来访问数据库对象，完成数据库中的数据和 Java 对象里的一种关联关系的一系列操作 CRUD。数据库中的对象有很多，每一个对象都写一个 DAO，显得很烦琐，每一个 DAO 都要写 CRUD 操作，这样代码的重复率高，如果使用泛型，代码的复用得到了很好的应用，提高了代码的效率。

4.2 相关知识

4.2.1 泛型的概念

所谓“泛型”，就是“宽泛的数据类型”，任意的数据类型。泛型是 Java 中一个非常重要的知识点，在 Java 集合类框架中泛型被广泛应用。使用泛型可以很好地解决“代码复用”问题。

4.2.2 泛型的定义和使用

1. 定义泛型类

在定义带类型参数的类时，在紧跟类名之后的 <> 内，指定一个或多个类型参数的名字，同时也可以对类型参数的取值范围进行限定，多个类型参数之间用“,”号分隔。

定义完类型参数后，可以在定义位置之后的类的几乎任意地方（静态块、静态属性、静态方法除外）使用类型参数，就像使用普通的类型一样。注意，父类定义的类型参数不能被子类继承。

```
public class TestClassDefine<T, S extends T> {
    …
}
```

2. 泛型方法

在定义带类型参数的方法时，在紧跟可见范围修饰（例如 public）之后的 <> 内，指定一个或多个类型参数的名字，同时也可以对类型参数的取值范围进行限定，多个类型参数之间用“,”号分隔。

定义完类型参数后，可以在定义位置之后的方法的任意地方使用类型参数，就像使用普通的类型一样。例如：

```
public <T, S extends T> T testGenericMethodDefine(T t, S s){
    ...
}
```

3. 泛型接口

先定义泛型接口：

```
public interface  Generator<T>{
    public T next();
}
```

然后定义这个实现类来实现这个接口：

```
public class GeneratorImpl  implements Generator<String>{
    @Override
    public String next(){
        ...
    }
}
```

4.2.3 相关概念

1. 通配符

类型通配符一般是使用“?”代替具体的类型参数，对类型参数赋予不确定值。例如 List<?> 在逻辑上是 List<String>、List<Integer> 等所有 List< 具体类型实参 > 的父类。

```
public class GenericTest {

    public static void main(String[] args) {
        List<String> name = new ArrayList<String>();
        List<Integer> age = new ArrayList<Integer>();
        List<Number> number = new ArrayList<Number>();

        name.add("icon");
        age.add(18);
        number.add(314);

        getData(name);
        getData(age);
        getData(number);

    }

   public static void getData(List<?> data) {
      System.out.println("data :" + data.get(0));
   }
}
```

运行结果：

```
data :icon
data :18
data :314
```

2. 上下边界

如果想限制使用泛型类别时，只能用某个特定类型或者是其子类型才能实例化该类型时，可以在定义类型时，使用 extends 关键字指定这个类型必须是继承某个类，或者实现某个接口，也可以是这个类或接口本身。

（1）类型通配符上限通过 List 来定义，如此定义就是通配符泛型值接受 Number 及其下层子类类型。

```
public class GenericTest {

    public static void main(String[] args) {
        List<String> name = new ArrayList<String>();
        List<Integer> age = new ArrayList<Integer>();
        List<Number> number = new ArrayList<Number>();

        name.add("icon");
        age.add(18);
        number.add(314);

        //getUperNumber(name);//1
        getUperNumber(age);//2
        getUperNumber(number);//3

    }

    public static void getData(List<?> data) {
       System.out.println("data :" + data.get(0));
    }

    public static void getUperNumber(List<? extends Number> data) {
          System.out.println("data :" + data.get(0));
       }
}
```

解析：在 //1 处会出现错误，因为 getUperNumber() 方法中的参数已经限定了参数泛型上限为 Number，所以泛型为 String 不在这个范围之内，所以会报错。

（2）类型通配符下限通过形如 List<? extends Number> 来定义，表示类型只能接受 Number 及其三层父类类型，如 Objec 类型的实例。

3. 擦除

Java 中的泛型基本上都是在编译器这个层次来实现的。生成的 Java 字节码中，是不包含泛型中的类型信息的。使用泛型的时候加上的类型参数，编译器在编译的时候去掉，这个过程就称为类型擦除。

如在代码中定义 List<Integer> 和 List<String> 等类型，在编译后都会编成 List。JVM 看到的只是 List，而由泛型附加的类型信息对 JVM 来说是不可见的。

4.2.4 泛型的好处

（1）类型安全。

通过知道使用泛型定义的变量的类型限制，编译器可以更有效地提高 Java 程序的类型安全。

（2）消除强制类型转换。

消除源代码中的许多强制类型转换。这使得代码可读性更强，并且减少了出错机会。所有的强制转换都是自动和隐式的。

（3）提高性能。

```
Lits list1 = new ArrayList();
list1.add("CSDN_SEU_Cavin ");
```

```
String str1 = (String)list1.get(0);
List<String> list2 = new ArrayList<String>();
list2.add("CSDN_SEU_Cavin ");  String str2 = list2.get(0);
```

对于上面的两段程序，由于泛型所有工作都在编译器中完成，Javac 编译出来的字节码是一样的（只是更能确保类型安全），那么何谈性能提升呢？是因为在泛型的实现中，编译器将强制类型转换插入生成的字节码中，但是更多类型信息可用于编译器这一事实，为未来版本的 JVM 的优化带来了可能。

4.2.5 泛型使用时的注意事项

（1）泛型的类型参数只能是类类型（包括自定义类），不能是简单类型。

（2）泛型的类型参数可以有多个。

4.3 任务实施

任务一 泛型类的定义和使用

1. 任务需求

先来看没有泛型的情况下的容器类如何定义：

```
package com.daiinfo.seniorjava.ken4.implement;
public class Boxs {
    private String key;
    private String value;
    public Boxs(String k, String v) {
        key = k;
        value = v;
    }

    public String getKey() {
        return key;
    }
    public void setKey(String key) {
        this.key = key;
    }
    public String getValue() {
        return value;
    }
    public void setValue(String value) {
        this.value = value;
    }
}
```

Box 类保存了一对 key-value 键值对，但是类型是定死的，也就说如果我想要创建一个 String-Integer 类型的键值对，当前这个 Box 是做不到的，还必须要另外重写一个 Box，这明显重用性就非常低，代码得不到复用，使用泛型可以很好地解决这个问题。

2. 任务分析

重新定义 Box 类，类图如图 4-1 所示。

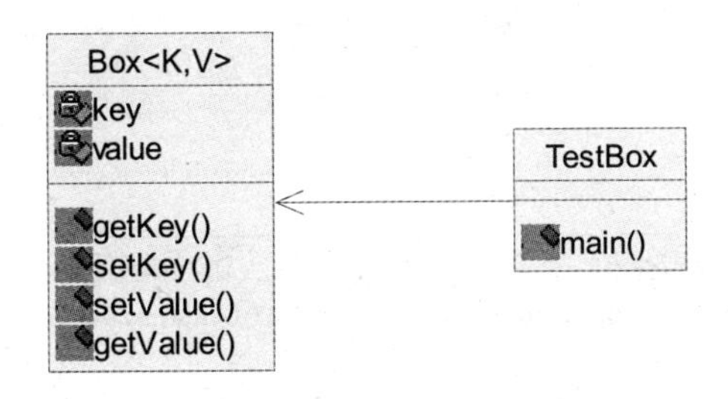

图4-1　类图

3. 任务实现

Box.java

```
package com.daiinfo.seniorjava.ken4.implement;
public class Box<K, V> {
    private K key;
    private V value;
    public Box(K k, V v) {
        key = k;
        value = v;
    }
    public K getKey() {
        return key;
    }
    public void setKey(K key) {
        this.key = key;
    }
    public V getValue() {
        return value;
    }
    public void setValue(V value) {
        this.value = value;
    }
}
```

在编译期是无法知道 K 和 V 具体是什么类型的，只有在运行时才会真正根据类型来构造和分配内存。这样我们的 Box 类便可以得到复用，我们可以将 T 替换成任何我们想要的类型。实例化泛型类的时候，我们只需要把类型参数换成具体的类型即可。可以看一下现在 Box 类对于不同类型的支持情况。

TestBox.java

```
package com.daiinfo.seniorjava.ken4.implement;
public class TestBox {
    public static void main(String[] args) {
        // TODO Auto-generated method stub
        Box<String, String> c1 = new Box<String, String>("name", "findingsea");
        Box<String, Integer> c2 = new Box<String, Integer>("age", 24);
        Box<Double, Double> c3 = new Box<Double, Double>(1.1, 2.2);
        System.out.println(c1.getKey() + " : " + c1.getValue());
```

```
        System.out.println(c2.getKey() + " : " + c2.getValue());
        System.out.println(c3.getKey() + " : " + c3.getValue());
    }
}
```

运行结果如图 4-2 所示。

```
Markers  Properties  Servers  Data Source Explorer  Snippets  Console ⊠  Search
<terminated> TestBox (1) [Java Application] C:\Java\jdk1.7.0_67\bin\javaw.exe (2017年7月31日 下午12:15:52)
name : findingsea
age : 24
1.1 : 2.2
```

图 4-2　运行结果

通过public class Container<K, V> {}定义泛型类，在实例化该类时，必须指明泛型K、V的具体类型，例如：Container<String, String> c1 = new Container<String, String>("name", "Messi");，指明泛型K的类型为String，泛型V的类型为String。

任务二　泛型方法的定义和使用

1. 任务需求

定义和使用泛型方法。

2. 任务分析

声明一个泛型方法很简单，只要在返回类型前面加上一个类似 <K，V> 的形式就行了。其类图如图 4-3 所示。

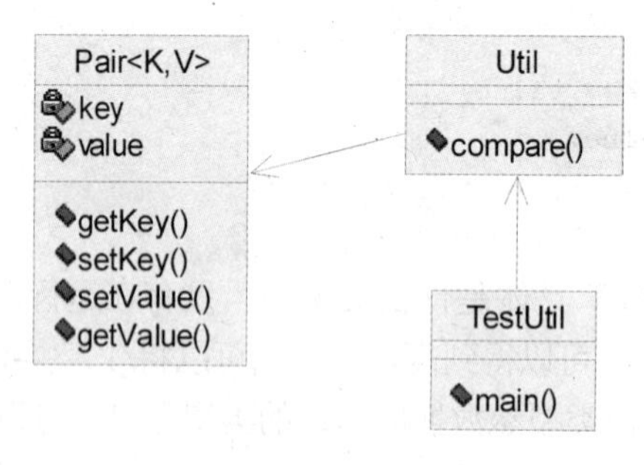

图4-3　类图

3. 任务实现

Pair.java

```
package com.daiinfo.seniorjava.ken4.implement;
public class Pair<K,V> {
    private K key;
    private V value;

    public Pair(K key, V value) {
```

```
        this.key = key;
        this.value = value;
    }

    public K getKey() {
        return key;
    }
    public V getValue() {
        return value;
    }

    public void setKey(K key) {
        this.key = key;
    }
    public void setVale(V value) {
        this.value = value;
    }
}
```

Util.java

```
package com.daiinfo.seniorjava.ken4.implement;
public class Util {
    public static <K, V> boolean compare(Pair<K, V> p1, Pair<K, V> p2) {
        return p1.getKey().equals(p2.getKey())&&p1.getValue().equals(p2.getValue());
    }
}
```

TestUtil.java

```
package com.daiinfo.seniorjava.ken4.implement;
public class TestPair {
    public static void main(String[] args) {
        // TODO Auto-generated method stub
        Pair<String, String> p1=new Pair<String, String>("name", "zhang");
        Pair<String, String> p2=new Pair<String, String>("name", "liu");
        System.out.println("比较结果: p1=p2\t"+Util.compare(p1, p2));

        Pair<String, Integer> p3=new Pair<String, Integer>("age", 67);
        Pair<String, Integer> p4=new Pair<String, Integer>("age", 67);
        System.out.println("比较结果: p3=p4\t"+Util.compare(p3, p4));

        Pair<Integer, String> p5 = new Pair<Integer, String>(1, "apple");
        Pair<Integer, String> p6 = new Pair<Integer, String>(2, "pear");
        boolean same = Util.compare(p1, p2);
        System.out.print("比较结果:p5=p6\t"+same);
    }
}
```

运行结果如图 4-4 所示。

```
Markers  Properties  Servers  Data Source Explorer  Snippets  Console  Search
<terminated> TestPair [Java Application] C:\Java\jdk1.7.0_67\bin\javaw.exe (2017年8月3日 上午12:21:37)
比较结果：p1=p2    false
比较结果：p3=p4    true
比较结果:p5=p6    false
```

图4-4　运行结果

在调用泛型方法的时候，在不指定泛型的情况下，泛型变量的类型为该方法中的几种类型的同一个父类的最小级，直到Object。在指定泛型的时候，该方法中的几种类型必须是该泛型实例类型或者其子类。

任务三　泛型接口的定义和使用

1. 任务需求

对任两个数求和。这两个数可以是整数、浮点数和字符串。

2. 任务分析

定义泛型接口 Calculator，其中定义泛型方法 and，用来求和。

3. 任务实现

```java
package com.daiinfo.seniorjava.ken4.implement;
public interface Calculator<T> {
    public T and(T a, T b);
}
```

定义类 CalculatorInteger 实现 Calculator 接口。

```java
package com.daiinfo.seniorjava.ken4.implement;
public class CalculatorInteger implements Calculator<Integer> {
    @Override
    public Integer and(Integer a, Integer b) {
        // TODO Auto-generated method stub
        return a+b;
    }
}
```

定义类 CalculatorString 实现 Calculator 接口。

```java
package com.daiinfo.seniorjava.ken4.implement;
public class CalculatorString implements Calculator<String>{
    @Override
    public String and(String a, String b) {
        // TODO Auto-generated method stub
        return a+b;
    }
}
```

定义测试类 TestCalculator，进行测试。

```
package com.daiinfo.seniorjava.ken4.implement;
public class TestCalculator {
    public static void main(String[] args) {
        // TODO Auto-generated method stub
        CalculatorInteger ci = new CalculatorInteger();
        Integer val = ci.and(10, 20);
        System.out.println(val);

        CalculatorString cs = new CalculatorString();
        String string = cs.and("湖北", "武汉");
        System.out.println(string);
    }
}
```

运行结果如图 4-5 所示。

```
Markers  Properties  Servers  Data Source Explorer  Snippets  Console  Search
<terminated> TestCalculator [Java Application] C:\Java\jdk1.7.0_67\bin\javaw.exe (2017年8月3日 上午12:06:40)
30
湖北武汉
```

图4-5　运行结果

通过public interface Calculator<T> {}定义泛型接口，在实现该接口时，必须指明泛型T的具体类型，例如：public class CalculatorInteger implements Calculator<Integer>{}，指明泛型T的类型为Integer，实例化该类时无须指定泛型类型。

4.4　拓展知识

在开发高校图书管理系统时，涉及多个数据库中的表，如：图书基本信息表 Book、教师信息表 Teacher、学生信息表 Student、用户表 User、出版社信息表 Publishing、图书分类信息表 Category 等。通过 DAO（Date Access Object，数据访问对象）实现对数据库中的表进行数据操作，就是数据库中的数据和 Java 对象里的一种关联关系的一系列操作，包括增删改查（CRUD）操作。

这些基本信息表对应 Java 的实体类。每个实体类都包含 CRUD 操作。

```
public class UserDao {
    void add(User user);
    void delete(User user);
    void update(User user);
    User select(User user);
}
```

```
public class TeacherDao {
    void add(Teacher tea);
```

```
    void delete(Teacher tea);
    void update(Teacher tea);
    User select(Teacher tea);
}
```

这样使得代码冗余度较高。那么能否将这多个 DAO 相同的方法封装成一个呢？这是可以的，即使用泛型类和泛型方法。抽象一个 BaseDao，里面封装了实体类相同的操作，当需要操作表的时候，将 T 换成 User 或者是 Teacher 就可以了。

4.5 拓展训练

1. 任务需求

应用泛型简单的 CURD 的实现。

2. 任务分析

设计 Dao（数据访问层），Bean（实体类），Impl（具体方法实现），Test（测试类），JDBC（数据库连接层），通过相对应的程序类设计结合泛型实现简单的增、删、改、查。

这些类的包结构如图 4-6 所示。

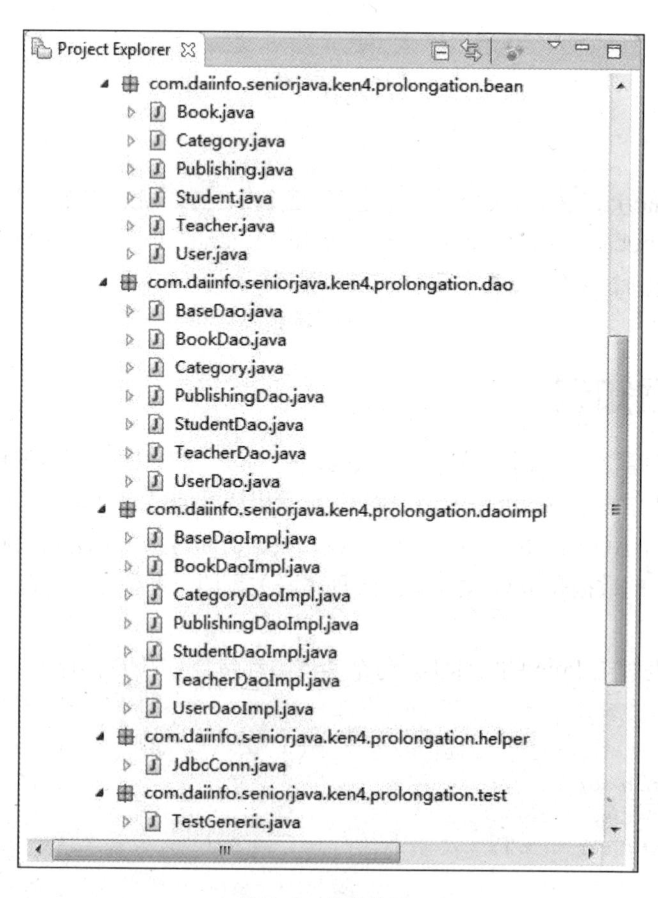

图4-6　包结构图

这些类的类图如图 4-7 所示。

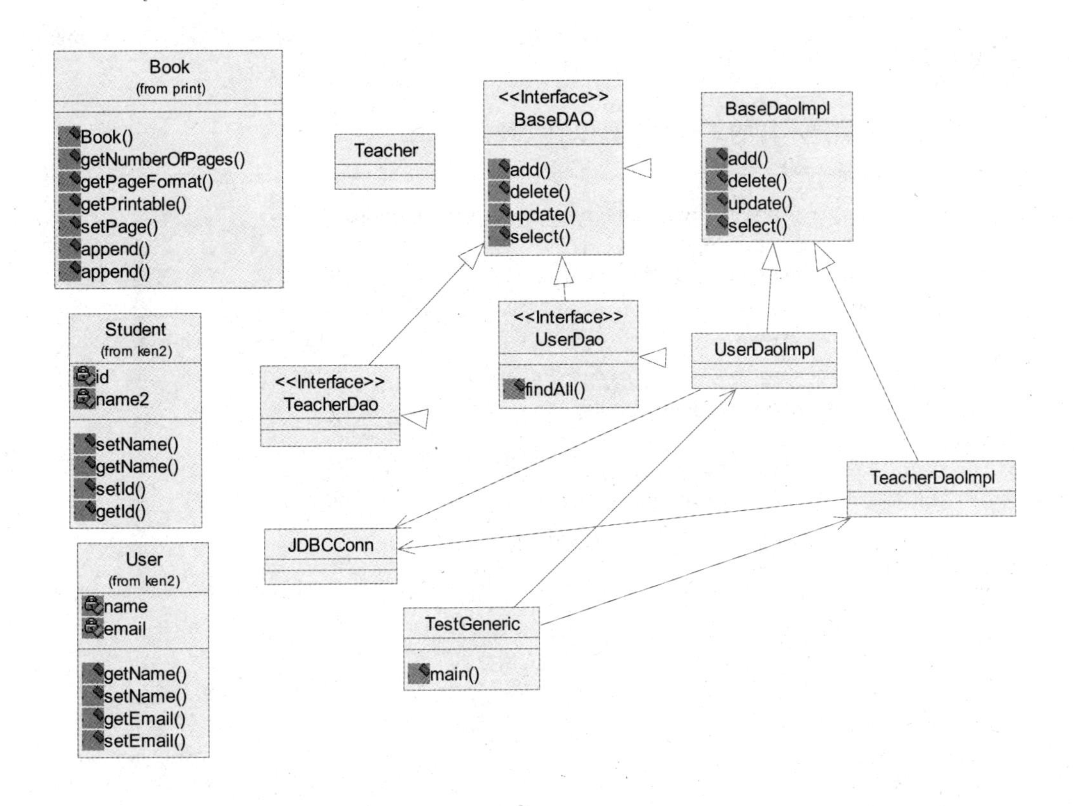

图4-7　类图

3. 任务实现

BaseDao.java

```
package com.daiinfo.seniorjava.ken4.prolongation.dao;
public interface BaseDao<T> {
    void add(T t);
    void delete(T t);
    void update(T t);
    T select(T t);
}
```

Teacher.java

```
package com.daiinfo.seniorjava.ken4.prolongation.dao;
public interface TeacherDao<T> extends BaseDao<T> {
}
```

这样我们只需要定义一个新的 UserDao 就可以了，它继承了 BaseDao 中的所有方法，当有额外的业务逻辑的时候，只需要添加额外方法就可以了。

UserDao.java

```
package com.daiinfo.seniorjava.ken4.prolongation.dao;
import java.util.List;
import com.hbliti.example4.generic.bean.User;
public interface UserDao extends BaseDao<User> {
```

```
    //查询所有的用户
    List<User>findAll();
}
```

这样光有接口当然是不行的，需要有具体的实现类。

BaseDaoImpl.java

```
package com.daiinfo.seniorjava.ken4.prolongation.daoimpl;
import java.lang.reflect.Field;
import java.lang.reflect.ParameterizedType;
import java.sql.PreparedStatement;
import java.sql.ResultSet;
import com.hbliti.example4.generic.dao.BaseDao;
import com.hbliti.example4.generic.helper.JdbcConn;;
//通用DAO
public class BaseDaoImpl<T> implements BaseDao<T> {
    /** 操作常量 */
    public static final String SQL_INSERT = "insert";
    public static final String SQL_UPDATE = "update";
    public static final String SQL_DELETE = "delete";
    public static final String SQL_SELECT = "select";
    private Class<T> EntityClass; // 获取实体类
    private PreparedStatement statement;
    private String sql;
    private Object argType[];
    private ResultSet rs;
    @SuppressWarnings("unchecked")
    public BaseDaoImpl() {

        /**
         *  传递User就是 com.example.daoimp.BaseDaoImp<com.example.bean.User>
         *  传递Shop就是 com.example.daoimp.BaseDaoImp<com.example.bean.Shop>
         * */
        ParameterizedType type = (ParameterizedType) getClass()
                .getGenericSuperclass();

        /**
         * 这里如果传递的是User，那么就是class com.example.bean.User
         * 如果传递的是Shop，那么就是class com.example.bean.Shop
         * */
        EntityClass = (Class<T>) type.getActualTypeArguments()[0];
    }
    @Override
    public void add(T t) {
        // TODO Auto-generated method stub
        sql = this.getSql(SQL_INSERT);   //获取sql
        // 赋值
        try {
            argType = setArgs(t, SQL_INSERT);
            statement = JdbcConn.getPreparedStatement(sql);  //实例化PreparedStatement
            //为sql语句赋值
            statement = JdbcConn.setPreparedStatementParam(statement,
```

```
                argType);
            statement.executeUpdate(); //执行语句
        } catch (Exception e) {
            // TODO Auto-generated catch block
            e.printStackTrace();
        } finally {
            JdbcConn.release(statement, null);  //释放资源
        }
    }
    @Override
    public void delete(T t) {
        // TODO Auto-generated method stub
        sql = this.getSql(SQL_DELETE);
        try {
            argType = this.setArgs(t, SQL_DELETE);
            statement = JdbcConn.getPreparedStatement(sql);
            statement = JdbcConn.setPreparedStatementParam(statement,
                    argType);
            statement.executeUpdate();
        } catch (Exception e) {
            // TODO Auto-generated catch block
            e.printStackTrace();
        } finally {
            JdbcConn.release(statement, null);
        }
    }
    @Override
    public void update(T t) {
        // TODO Auto-generated method stub
        sql = this.getSql(SQL_UPDATE);
        try {
            argType = setArgs(t, SQL_UPDATE);
            statement = JdbcConn.getPreparedStatement(sql);
            statement = JdbcConn.setPreparedStatementParam(statement,
                    argType);
            statement.executeUpdate();
        } catch (Exception e) {
            // TODO Auto-generated catch block
            e.printStackTrace();
        } finally {
            JdbcConn.release(statement, null);
        }
    }
    @Override
    public T select(T t) {
        // TODO Auto-generated method stub
        sql = this.getSql(SQL_SELECT);
        T obj = null;
        try {
            argType = setArgs(t, SQL_SELECT);
            statement = JdbcConn.getPreparedStatement(sql);
```

```
            statement = JdbcConn.setPreparedStatementParam(statement,
                    argType);
            rs = statement.executeQuery();
            Field fields[] = EntityClass.getDeclaredFields();
            while (rs.next()) {
                obj = EntityClass.newInstance();
                for (int i = 0; i < fields.length; i++) {
                    fields[i].setAccessible(true);
                    fields[i].set(obj, rs.getObject(fields[i].getName()));
                }
            }
        } catch (Exception e) {
            // TODO Auto-generated catch block
            e.printStackTrace();
        }
        return obj;
    }
/**
* SQL拼接函数
*形如 : insert into User(id,username,password,email,grade) values(?,?,?,?,?)
**/
    private String getSql(String operator) {
        StringBuffer sql = new StringBuffer();
        // 通过反射获取实体类中的所有变量
        Field fields[] = EntityClass.getDeclaredFields();
        // 插入操作
        if (operator.equals(SQL_INSERT)) {
            sql.append("insert into " + EntityClass.getSimpleName());
            sql.append("(");
            for (int i = 0; fields != null && i < fields.length; i++) {
                fields[i].setAccessible(true);     //这句话必须要有,否则会抛出异常
                String column = fields[i].getName();
                sql.append(column).append(",");
            }
            sql = sql.deleteCharAt(sql.length() - 1);
            sql.append(") values (");
            for (int i = 0; fields != null && i < fields.length; i++) {
                sql.append("?,");
            }
            sql.deleteCharAt(sql.length() - 1);
            // 是否需要添加分号
            sql.append(")");
        } else if (operator.equals(SQL_UPDATE)) {
            sql.append("update " + EntityClass.getSimpleName() + " set ");
            for (int i = 0; fields != null && i < fields.length; i++) {
                fields[i].setAccessible(true);
                String column = fields[i].getName();
                if (column.equals("id")) {
                    continue;
                }
                sql.append(column).append("=").append("?,");
```

```
            }
            sql.deleteCharAt(sql.length() - 1);
            sql.append(" where id=?");
        } else if (operator.equals(SQL_DELETE)) {
            sql.append("delete from " + EntityClass.getSimpleName()
                    + " where id=?");
        } else if (operator.equals(SQL_SELECT)) {
            sql.append("select * from " + EntityClass.getSimpleName()
                    + " where id=?");
        }
        return sql.toString();
    }
    // 获取参数
    private Object[] setArgs(T entity, String operator)
            throws IllegalArgumentException, IllegalAccessException {
        Field fields[] = EntityClass.getDeclaredFields();
        if (operator.equals(SQL_INSERT)) {
            Object obj[] = new Object[fields.length];
            for (int i = 0; obj != null && i < fields.length; i++) {
                fields[i].setAccessible(true);
                obj[i] = fields[i].get(entity);
            }
            return obj;
        } else if (operator.equals(SQL_UPDATE)) {
            Object Tempobj[] = new Object[fields.length];
            for (int i = 0; Tempobj != null && i < fields.length; i++) {
                fields[i].setAccessible(true);
                Tempobj[i] = fields[i].get(entity);
            }
            Object obj[] = new Object[fields.length];
            System.arraycopy(Tempobj, 1, obj, 0, Tempobj.length - 1);
            obj[obj.length - 1] = Tempobj[0];
            return obj;
        } else if (operator.equals(SQL_DELETE)) {
            Object obj[] = new Object[1];
            fields[0].setAccessible(true);
            obj[0] = fields[0].get(entity);
            return obj;
        } else if (operator.equals(SQL_SELECT)) {
            Object obj[] = new Object[1];
            fields[0].setAccessible(true);
            obj[0] = fields[0].get(entity);
            return obj;
        }
        return null;
    }
}
```

这样就对 BaseDao 进行了具体的实现。因为 User 表还有其他额外的操作，它通过继承 BaseDaoImpl，然后实现 UserDao 接口，那么 UserDaoImpl 就具有了 BaseDaoImpl 的通用方法，还具有了自己其他的额外方法。

UserDaoImpl.java

```
package com.daiinfo.seniorjava.ken4.prolongation.daoimpl;
import java.lang.reflect.ParameterizedType;
import java.sql.PreparedStatement;
import java.sql.ResultSet;
import java.util.ArrayList;
import java.util.List;
import com.hbliti.example4.generic.bean.User;
import com.hbliti.example4.generic.dao.UserDao;
import com.hbliti.example4.generic.helper.JdbcConn;
public class UserDaoImpl extends BaseDaoImpl<User> implements UserDao {
    private Class<?> EntityClass;
    private String sql;
    private PreparedStatement statement;
    private ResultSet rs;
    private List<User> list;
    public UserDaoImpl() {
        ParameterizedType type = (ParameterizedType) getClass()
                .getGenericSuperclass();
        EntityClass = (Class<?>) type.getActualTypeArguments()[0];
    }
    @Override
    public List<User> findAll() {
        // TODO Auto-generated method stub
        StringBuffer b = new StringBuffer();
        list = new ArrayList<User>();
        sql = b.append("select * from " + EntityClass.getSimpleName())
                .toString();
        try {
            statement = JdbcConn.getPreparedStatement(sql);
            rs = statement.executeQuery();
            while (rs.next()) {
                User user = new User();
                user.setId(rs.getInt("id"));
                user.setPassword(rs.getString("password"));
                user.setEmail(rs.getString("email"));
                user.setUsername(rs.getString("username"));
                user.setGrade(rs.getInt("grade"));
                list.add(user);
            }
        } catch (Exception e) {
            // TODO Auto-generated catch block
            e.printStackTrace();
        }
        return list;
    }
}
```

如果还有 Teacher 表，那么同理可以去创建一个 TeacherDao 去继承 BaseDao，然后在自己的 TeacherDao 中定义其他的额外方法就可以了。当表非常多的时候，就可以采用这种思想进行封装。这样写出的代码质量就显得非常的高，耦合度也非常的小。

再添加上工具类连接数据库类 JdbcConn。

JdbcConn.java

```
package com.daiinfo.seniorjava.ken4.prolongation.helper;
import java.sql.Connection;
import java.sql.DriverManager;
import java.sql.PreparedStatement;
import java.sql.ResultSet;
import java.sql.SQLException;
public class JdbcConn {
    private static final String USER = "root";
    private static final String PASSWORD = "123456";
    private static final String URL = "jdbc:mysql://localhost:3306/usermanager";
    private static Connection con;
    // 获取数据库连接对象
    public static Connection getConnection() {
        if (con == null) {
            try {
                Class.forName("com.mysql.jdbc.Driver");
                con = DriverManager.getConnection(URL, USER, PASSWORD);
            } catch (ClassNotFoundException e) {
                // TODO Auto-generated catch block
                e.printStackTrace();
            } catch (SQLException e) {
                // TODO Auto-generated catch block
                e.printStackTrace();
            }
        } else {
            return con;
        }
        return con;
    }
    public static PreparedStatement getPreparedStatement(String sql)
            throws SQLException {
        return getConnection().prepareStatement(sql);
    }
    public static PreparedStatement setPreparedStatementParam(
            PreparedStatement statement, Object obj[]) throws SQLException {
        for (int i = 0; i < obj.length; i++) {
            statement.setObject(i + 1, obj[i]);
        }
        return statement;
    }
    // 释放资源
    public static void release(PreparedStatement ps, ResultSet rs) {
        try {
            if (con != null) {
                con.close();
                con = null;
            }
            if (ps != null) {
                ps.close();
```

```
                ps = null;
            }
            if (rs != null) {
                rs.close();
                rs = null;
            }
        } catch (Exception e) {
            // TODO: handle exception
        }
    }
}
```

User.java

```
package com.daiinfo.seniorjava.ken4.prolongation.bean;
public class User {
    private int id;
    private String username;
    private String password;
    private String email;
    private int grade;

    public User(){
    }

    public User(int id,String username,String password,String email,int grade){
        this.id = id;
        this.username = username;
        this.password = password;
        this.email = email;
        this.grade = grade;
    }
    public int getId() {
        return id;
    }
    public void setId(int id) {
        this.id = id;
    }
    public String getUsername() {
        return username;
    }
    public void setUsername(String username) {
        this.username = username;
    }
    public String getPassword() {
        return password;
    }
    public void setPassword(String password) {
        this.password = password;
    }
    public String getEmail() {
        return email;
```

```
    }
    public void setEmail(String email) {
        this.email = email;
    }
    public int getGrade() {
        return grade;
    }
    public void setGrade(int grade) {
        this.grade = grade;
    }
}
```

测试类 TestGeneric 代码如下。

TestGeneric.java

```
package com.daiinfo.seniorjava.ken4.prolongation.test;
import java.util.List;
import com.hbliti.example4.generic.bean.User;
import com.hbliti.example4.generic.daoimpl.UserDaoImpl;
public class TestGeneric {
    /**
     * @param args
     */
    public static void main(String[] args) {
        // TODO Auto-generated method stub
        List<com.hbliti.example4.generic.bean.User> list = null;
        UserDaoImpl imp = new UserDaoImpl();
        list = imp.findAll();
        for (User user : list) {
            System.out.println(user.getId() + " " + user.getUsername() + " "
                    + user.getPassword() + " " + user.getEmail() + " "
                    + user.getGrade());
        }
        // insert操作
        User user = new User();
        user.setId(1);
        user.setUsername("张三");
        user.setPassword("123456");
        user.setEmail("zhangsan@sina.com ");
        user.setGrade(5);
        imp.add(user);
        // update操作
        // User user = new User();
        user.setId(1);
        user.setUsername("李四");
        user.setPassword("123456");
        user.setEmail("lisi@qq.com ");
        user.setGrade(4);
        imp.update(user);
    }
}
```

运行结果如图 4-8 所示。

图4-8 运行结果

4.6 课后小结

1. 泛型的本质便是类型参数化，通俗地说就是用一个变量来表示类型，这个类型可以是 String、Integer 等等不确定，表明可接受的类型。

2. 泛型可以接受多个参数，而 Object 经过强制类型转换可以转换为任何类型，泛型可以把使用 Object 的错误提前到编译后，而不是运行后，提升安全性。

3. 在 Java 的虚拟机中并不存在泛型，泛型只是为了完善 Java 体系，增加程序员编程的便捷性以及安全性而创建的一种机制，在 Java 虚拟机中对应泛型的都是确定的类型，在编写泛型代码后，Java 虚拟机中会把这些泛型参数类型都擦除，用相应的确定类型来代替，代替的这一动作叫作类型擦除，而用于替代的类型称为原始类型，在类型擦除过程中，一般使用第一个限定的类型来替换，若无限定则使用 Object。

4. 泛型限定是通过?（通配符）来实现的，表示可以接受任意类型。

5. 泛型的一些基本规则约束如下。

① 泛型的类型参数必须为类的引用，不能用基本类型（int、short、long、byte、float、double、char、boolean）。

② 泛型是类型的参数化，在使用时可以用作不同类型（此处在介绍泛型类时会详细说明）。

③ 泛型的类型参数可以有多个。

4.7 课后习题

一、填空题

1. Java 泛型可以使用 3 种通配符进行限制，分别是：________、extends、super。

2. List<? extends T>; 的作用是________。

3. public V put(K key, V value) {} 中，K 代表________，V 代表________。

二、选择题

1. 以下哪种书写是正确的________。

A. ArrayList<String> lists = new ArrayList<String>();

B. ArrayList<Object> lists = new ArrayList<String>();

C. ArrayList<String> lists = new ArrayList<Object>();

D. ArrayList lists = new ArrayList();

2. 泛型使用中的规则和限制是________。

A. 泛类参数只能是类类型不能是简单类型

B. 同一种泛型可以对应多个版本

C. 泛型的类型参数可以有多个

D. 以上都是

3. 下面关于泛型的说法不正确的是________。

A. 泛型的具体确定时间可以是在定义方法的时候

B. 泛型的具体确定时间可以是在创建对象的时候

C. 泛型的具体确定时间可以是在继承父类定义子类的时候

D. 泛型就是 Object 类型

4. 父类声明：public class FXfather<T>{…}

现在要定义一个 Fxfather 的子类 son，下面定义错误的是________。

A. class Son extends FXfather<String>{}

B. class Son<T,V> extends FXfather<T>{}

C. class Son<String> extends FXfather<String>{}

D. class Son<String> extends FXfather<T>{}

5. 关于泛型的说法正确的是________。

A. 泛型是 JDK 1.5 出现的新特性　　B. 泛型是一种安全机制

C. 使用泛型避免了强制类型转换　　D. 使用泛型必须进行强制类型转换

三、简答题

1. Java 中的泛型是什么？使用泛型的好处是什么？

2. Java 的泛型是如何工作的？什么是类型擦除？

3. 什么是泛型中的限定通配符和非限定通配符？

4. 如何编写一个泛型方法，让它能接受泛型参数并返回泛型类型？

4.8 上机实训

实训　假如我们现在要定义一个类来表示坐标，要求坐标的数据类型可以是整数、小数和字符串，例如：

- x = 10、y = 10
- x = 12.88、y = 129.65
- x = "东京 180 度"、y = "北纬 210 度"

定义泛型类 Point，其中定义一个打印坐标的泛型方法 printPoint()。并定义一个测试类进行测试，以不同的格式输出给定的坐标值。

PART05

知识领域5 Java序列化机制

知识目标

1. 掌握对象序列化的基本概念。
2. 掌握对象序列化的方法。

■ 能力目标

熟练使用对象的序列化编写相应应用程序。

■ 素质目标

1. 能够阅读科技文档和撰写分析文档。
2. 能够查阅JDK API。
3. 增强学生团队协作能力。

5.1 应用场景

在分布式环境下，当进行远程通信时，彼此可以发送各种类型的数据。无论是何种类型的数据，都会以二进制序列的形式在网络上传输。发送方需要把这个 Java 对象转换为字节序列，才能在网络上传送；接收方则需要把字节序列再恢复为 Java 对象。

序列化是一种将对象以一连串的字节描述的过程，用于解决在对对象流进行读写操作时所引发的问题。序列化可以将对象的状态写在流里进行网络传输，或者保存到文件、数据库等系统中，并在需要时把该流读取出来重新构造一个相同的对象。

5.2 相关知识

5.2.1 序列化的概念

将在内存中的各种对象的状态（也就是实例变量，不是方法）保存在磁盘中或者在网络中进行传输，并且可以把保存的对象状态再读出来。

将一个 Java 对象写入 IO 流；与此对应的，则是从 IO 流中恢复一个 Java 对象。

Java 提供这种保存对象状态的机制，就是序列化。

对象序列化是 Java 编程中的必备武器。

5.2.2 序列化应用

序列化通常应用在以下情景。

- 想把内存中的对象状态保存到一个文件中或者数据库中的时候；
- 想用套接字在网络上传送对象的时候；
- 想通过 RMI 传输对象的时候。

5.2.3 序列化的几种方式

在 Java 中用 Socket 传输数据时，数据类型往往比较难选择。可能要考虑带宽、跨语言、版本的兼容等问题。比较常见的做法有以下两种。

- 把对象包装成 JSON 字符串传输；
- 采用 Java 对象的序列化和反序列化。

随着 Google 工具 protoBuf 的开源，protobuf 也成为不错的选择。

* 提示：对 JSON,Object Serialize,ProtoBuf 做个对比。

5.2.4 对象实现机制

为了方便开发人员将 Java 对象进行序列化及反序列化，Java 提供了一套方便的 API 来支持。其中包括以下接口和类。

- Java.io.Serializable；
- Java.io.Externalizable；
- ObjectOutput；
- ObjectInput；

- ObjectOutputStream；
- ObjectInputStream。

5.3 任务实施

任务一 使用 Serializable 序列化实体对象

1. 任务需求

将一个对象序列化后写入到本地文件中。

2. 任务分析

本任务类图如图 5-1 所示。

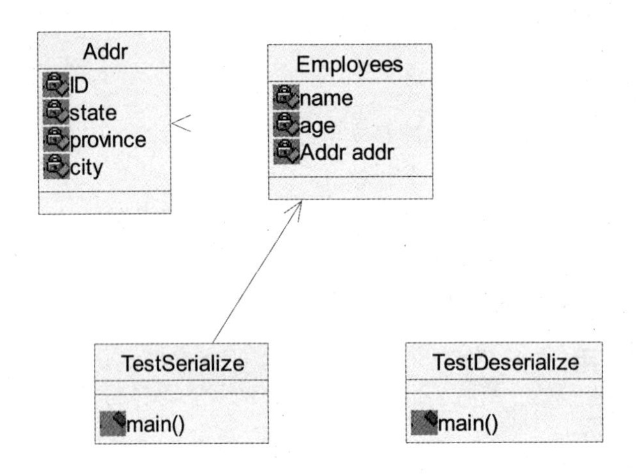

图5-1 类图

Employee.Java 类是一个可序列化的实体类，要实现 Serializable 接口，同时需要有一个成员变量 serialVersionUID。Address.Java 类描述员工的地址信息。TestSerialize.Java 类是测试类。文件是：D:\employee.dat。

实现 Serializable 接口非常简单，只要让 Java 实现 Serializable 接口即可，无须实现任何方法。

一个类一旦实现了 Serializable 接口，那么该类的对象就是可序列化的。实现类的对象的序列化可以使用 ObjectOutputStream，实现步骤如下。

- 创建 ObjectOutputStream 对象；
- 调用 ObjectOutputStream 的 writeObject 方法输出对象。

3. 任务实现

（1）地址类 Address.Java。

```
package com.daiinfo.seniorjava.ken5.implement;
import java.io.Serializable;
public class Address implements Serializable {
    private static final long serialVersionUID = 4983187287403615604L;
    private String state; // 表示员工所在的国家
    private String province; // 表示员工所在的省
```

```
    private String city; // 表示员工所在的市
    public Address(String state, String province, String city) {// 利用构造方法初始化各个域
        this.state = state;
        this.province = province;
        this.city = city;
    }
    public String getState() {
        return state;
    }
    public void setState(String state) {
        this.state = state;
    }
    public String getProvince() {
        return province;
    }
    public void setProvince(String province) {
        this.province = province;
    }
    public String getCity() {
        return city;
    }
    public void setCity(String city) {
        this.city = city;
    }
    public static long getSerialversionuid() {
        return serialVersionUID;
    }
    @Override
    public String toString() {// 使用地址属性表示地址对象
        StringBuilder sb = new StringBuilder();
        sb.append("国家: " + state + ", ");
        sb.append("省: " + province + ", ");
        sb.append("市: " + city);
        return sb.toString();
    }
}
```

（2）员工类 Employee.Java。

```
package com.daiinfo.seniorjava.ken5.implement;
import java.io.Serializable;
public class Employee implements Serializable {
    private static final long serialVersionUID = 3049633059823371192L;
    private String name; // 表示员工的姓名
    private int age; // 表示员工的年龄
    private Address address;// 表示员工的地址
    // 利用构造方法初始化各个域
    public Employee(String name, int age, Address address) {
        this.name = name;
        this.age = age;
        this.address = address;
    }
```

```
    public String getName() {
        return name;
    }
    public void setName(String name) {
        this.name = name;
    }
    public int getAge() {
        return age;
    }
    public void setAge(int age) {
        this.age = age;
    }
    public Address getAddress() {
        return address;
    }
    public void setAddress(Address address) {
        this.address = address;
    }
    @Override
    public String toString() {// 重写toString()方法
        StringBuilder sb = new StringBuilder();
        sb.append("姓名: " + name + ", ");
        sb.append("年龄: " + age + "\n");
        sb.append("地址: " + address);
        return sb.toString();
    }
}
```

（3）测试类 TestSerialize.Java。

```
package com.daiinfo.seniorjava.ken5.implement;
import java.io.File;
import java.io.FileInputStream;
import java.io.FileNotFoundException;
import java.io.FileOutputStream;
import java.io.IOException;
import java.io.ObjectInputStream;
import java.io.ObjectOutputStream;
public class TestSerialize {
    public static void main(String[] args) {
        // 创建一个员工对象
        Address address = new Address("中国", "吉林", "长春");// 创建address对象
        Employee employee = new Employee("张XX", 30, address);// 创建employee对象
        // 将该对象写入到文件中，序列化该对象
        File file = new File("d:\\employee23.dat");
        try {
            FileOutputStream fos = new FileOutputStream(file);
            ObjectOutputStream oos = new ObjectOutputStream(fos);
            oos.writeObject(employee);
            oos.flush();
            oos.close();
            fos.close();
```

```
        } catch (FileNotFoundException e) {
            // TODO Auto-generated catch block
            e.printStackTrace();
        } catch (IOException e) {
            // TODO Auto-generated catch block
            e.printStackTrace();
        }
    }
}
```

任务二　使用反序列化将 Person 对象从磁盘上读出

1. 任务需求

使用反序列化将 Employee 对象从磁盘上读出并修改员工信息，然后再写入到文件中。

2. 任务分析

相应的反序列化需要使用的类是 ObjectInputStream，反序列化的步骤如下。

- 创建 ObjectInputStream 对象；
- 使用 ObjectInputStream 的 readObject 方法取出对象。

3. 任务实现

重构代码，实现反序列化，代码如下。

TestDeserialize.java

```
package com.daiinfo.seniorjava.ken5.implement;
import java.io.File;
import java.io.FileInputStream;
import java.io.FileNotFoundException;
import java.io.FileOutputStream;
import java.io.IOException;
import java.io.ObjectInputStream;
import java.io.ObjectOutputStream;
public class TestDeserialize {
    public static void main(String[] args) {
        File file = new File("d:\\employee23.dat");
        // 从文件中读取对象，反序列化该对象
        Employee employee1 = null;
        try {
            FileInputStream fis = new FileInputStream(file);
            ObjectInputStream ois = new ObjectInputStream(fis);
            employee1 = (Employee) ois.readObject();
            System.out.println("修改前员工的信息: ");
            System.out.println(employee1);// 输出employee对象
        } catch (FileNotFoundException e) {
            // TODO Auto-generated catch block
            e.printStackTrace();
        } catch (IOException e) {
            // TODO Auto-generated catch block
            e.printStackTrace();
        } catch (ClassNotFoundException e) {
            // TODO Auto-generated catch block
```

```
                e.printStackTrace();
            }
            // 修改该员工的相关信息
            employee1.getAddress().setState("中国");
            employee1.getAddress().setProvince("湖北省");
            employee1.getAddress().setCity("武汉市");
            employee1.setAge(21);
            // 将该员工对象再次写入文件中，进行序列化
            try {
                FileOutputStream fos = new FileOutputStream(file);
                ObjectOutputStream oos = new ObjectOutputStream(fos);
                oos.writeObject(employee1);
                oos.flush();
                oos.close();
                fos.close();
            } catch (FileNotFoundException e) {
                // TODO Auto-generated catch block
                e.printStackTrace();
            } catch (IOException e) {
                // TODO Auto-generated catch block
                e.printStackTrace();
            }
            // 再次从文件中读取对象，进行反序列化
            Employee employee2 = null;
            try {
                FileInputStream fis = new FileInputStream(file);
                ObjectInputStream ois = new ObjectInputStream(fis);
                employee2 = (Employee) ois.readObject();
                System.out.println("修改后员工的信息：");
                System.out.println(employee2);// 输出employee对象
            } catch (FileNotFoundException e) {
                // TODO Auto-generated catch block
                e.printStackTrace();
            } catch (IOException e) {
                // TODO Auto-generated catch block
                e.printStackTrace();
            } catch (ClassNotFoundException e) {
                // TODO Auto-generated catch block
                e.printStackTrace();
            }
        }
    }
```

运行结果如图 5-2 所示。

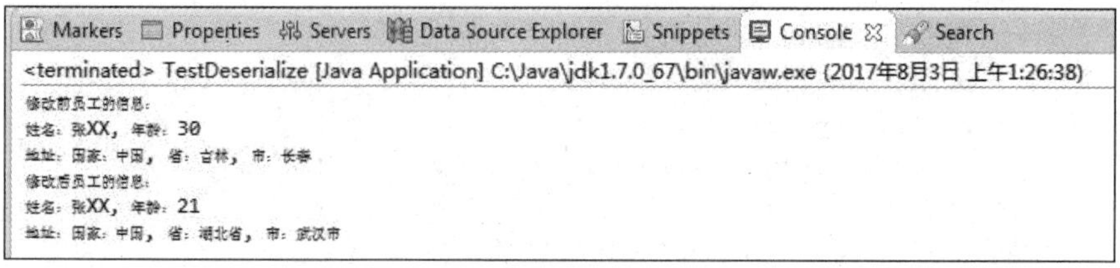

```
Markers  Properties  Servers  Data Source Explorer  Snippets  Console  Search
<terminated> TestDeserialize [Java Application] C:\Java\jdk1.7.0_67\bin\javaw.exe (2017年8月3日 上午1:26:38)
修改前员工的信息：
姓名：张XX，年龄：30
地址：国家：中国，省：吉林，市：长春
修改后员工的信息：
姓名：张XX，年龄：21
地址：国家：中国，省：湖北省，市：武汉市
```

图5-2　运行结果

关于对象序列化与反序列化还有几点需要注意。

- 反序列化无须通过构造器初始化对象；
- 如果使用序列化机制向文件中写入了多个对象，那么取出和写入的顺序必须一致；
- Java对类的对象进行序列化时，若类中存在对象引用（且值不为null），也会对类的引用对象进行序列化。

5.4 拓展知识

5.4.1 使用 transient

在一些特殊场景下，比如银行账户对象，出于保密考虑，不希望对存款金额进行序列化；或者类的一些引用类型的成员是不可序列化的。此时可以使用 transient 关键字修饰不想被或者不能被序列化的成员变量。

5.4.2 外部序列化

Java 语言还提供了另外一种方式来实现对象持久化，即外部序列化。外部序列化与序列化的主要区别在于序列化是内置的 API，只需要实现 Serializable 接口，开发人员不需要编写任何代码就可以实现对象的序列化，而用外部序列化时，Externalizable 接口中的方法必须有开发人员实现。因此与实现 Serializable 接口的方法相比，使用 Externalizable 编写程序的难度更大，但是由于控制权交给了开发者，在编程时有更多的灵活性，对需要持久化的那些属性可以进行控制，可能提高程序的性能。

5.5 拓展训练

1. 任务需求

用 Jackson 进行 JSON 解析和序列化。

2. 任务分析

常见 Java 序列化方式包括 Java 原生以流的方法进行的序列化、Json 序列化、FastJson 序列化、Protobuff 序列化。这里主要讲解 JSON 序列化。

Json（JavaScript Object Notation，JavaScript 对象表示方法），它是一个轻量级的数据交换格式，我们可以很简单地来读取和写它，并且它很容易被计算机转化和生成。

Json 的表现形式一般就 3 种（对象、数组、字符串），使用过程中，只是将对象与数组进行混合了。

（1）对象。

```
{
    "returnCode": "R0000",
    "returnMsg": "成功",
    "lessTimes": "2",
    "domainLink": "",
    "seqNum": "1",
```

```
    "identification": "595279",
    "isNeedImageCode": "false"
}
```

（2）数组。

```
{
    "employees": [
        { "firstName":"John" , "lastName":"Doe" },
        { "firstName":"Anna" , "lastName":"Smith" },
        { "firstName":"Peter" , "lastName":"Jones" }
    ]
}
```

（3）数组对象混合。

```
{
    "icon": [
        {
            "title": "上天猫，就购了",
            "icon_category": "baobei",
            "icon_key": "icon-service-tianmao",
            "trace": "srpservice",
            "innerText": "天猫宝贝",
            "url": "//www.tmall.com/"
        },
        {
            "title": "保险理赔",
            "position": "99",
            "innerText": "分组-保险理赔",
            "iconPopupComplex":
            {
                "popup_title": "保险理赔",
                "subIcons": [
                    {
                        "dom_class": "icon-service-yunfeixian",
                        "icon_content": "卖家赠送退货运费险"
                    }
                ]
            }
        }
    ]
}
```

对于比较复杂的 JSON 格式的字符串，可以在线验证 JSON 或使用 JSON 格式化工具，格式化一个格式，方便观看、整理、分析。

3．任务实现

Java 下常见的 JSON 类库有 Gson、JSON-lib 和 Jackson 等，Jackson 相对来说比较高效，在项目中主要使用 Jackson 进行 JSON 和 Java 对象转换，下面给出一些 Jackson 的 JSON 操作方法和步骤。

（1）准备工作。

首先下载 Jackson 工具包。Jackson 有 1.x 系列和 2.x 系列，2.x 系列有 3 个 jar 包需要下载。

- jackson-core-2.2.3.jar（核心 jar 包）；
- jackson-annotations-2.2.3.jar（该包提供 JSON 注解支持）；
- jackson-databind-2.2.3.jar。

（2）将 jar 包引入到工程。

① 右键项目，选择“Java Build Path”→“Libraries”选项卡，再单击“Add External JARs...”按钮，如图 5-3 所示。

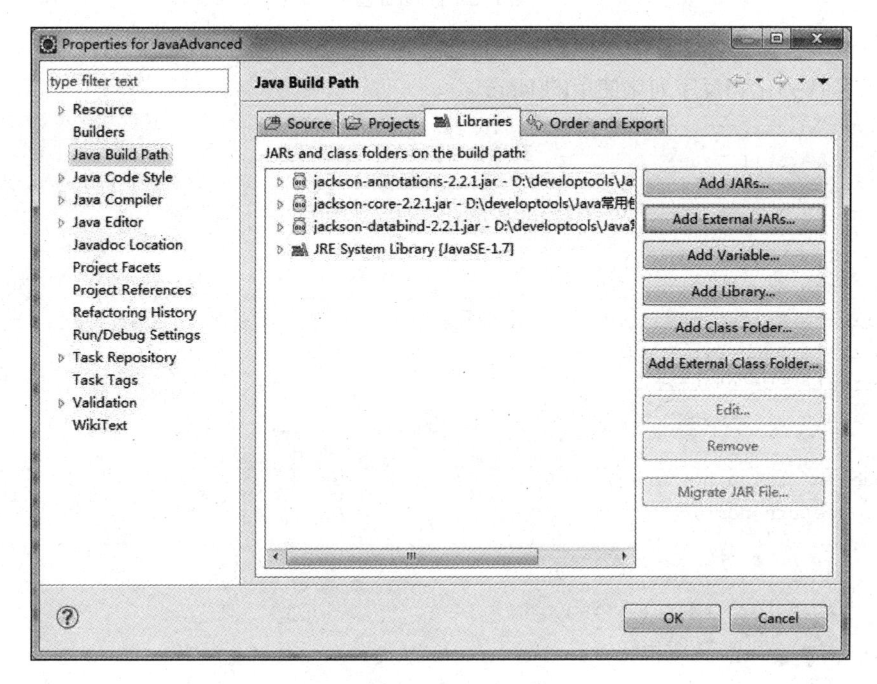

图5-3　编译路径

② 选择 3 个 jar 包文件，如图 5-4 所示。

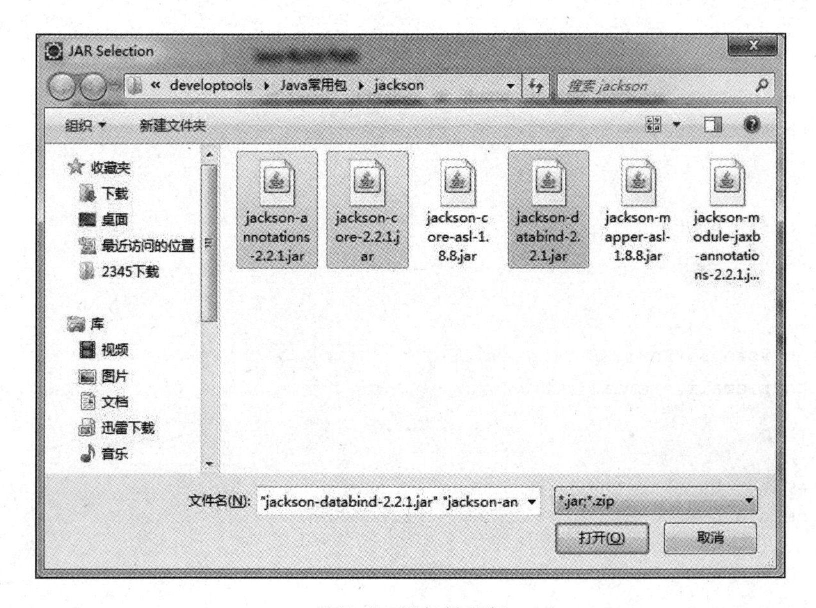

图5-4　添加扩展包

③ 引入完成，如图 5-5 所示。

JRE System Library [JavaSE-1.7]
jackson-annotations-2.2.1.jar - D:\developtools\Java常用包\ja
jackson-core-2.2.1.jar - D:\developtools\Java常用包\jackson
jackson-databind-2.2.1.jar - D:\developtools\Java常用包\jack

图5-5 扩展jar包

（3）新建项目和类。

① JSON 序列化和反序列化使用的 User 类。

```
package com.daiinfo.seniorjava.ken5.prolongation;
import java.util.Date;
import java.util.List;
import java.util.Set;
public class User {
    private String name;
    private String password;
    private Date birthday;
    private String email;
    private List<User> friends;

    public String getName() {
        return name;
    }

    public void setName(String name) {
        this.name = name;
    }

    public Date getBirthday() {
        return birthday;
    }

    public void setBirthday(Date birthday) {
        this.birthday = birthday;
    }

    public String getEmail() {
        return email;
    }

    public void setEmail(String email) {
        this.email = email;
    }

    public String getPassword() {
        return password;
    }

    public void setPassword(String password) {
```

```
            this.password = password;
        }

        public List<User> getFriends() {
            return friends;
        }

        public void setFriends(List<User> friends) {
            this.friends = friends;
        }

        public String toString(){
            return "姓名: "+name+"\n"+"密码: "+password+"\n"+"出生日期: "+birthday+"\n"+"邮箱:
"+email+""+"朋友"+friends;
        }
    }
```

② 测试类 TestJsonSerialize。

Java 对象转 JSON（JSON 序列化）和 JSON 转 Java 类（JSON 反序列化）。

```
    package com.daiinfo.seniorjava.ken5.prolongation;
    import java.io.IOException;
    import java.text.ParseException;
    import java.text.SimpleDateFormat;
    import java.util.ArrayList;
    import java.util.List;
    import com.fasterxml.jackson.databind.ObjectMapper;
    /**
     *
     * @author daiyuanquan
     *
     */
    public class TestJsonSerialize {
        public static void main(String[] args) throws IOException {
            try {
                new TestJsonSerialize().start();
            } catch (ParseException e) {
                // TODO Auto-generated catch block
                e.printStackTrace();
            }
        }
        public void start() throws IOException, ParseException {
            SimpleDateFormat formatter = new SimpleDateFormat("yyyyMMdd");

            //构建用户
            User u = new User();
            List<User> friends = new ArrayList<>();
            u.setName("张三");
            u.setPassword("123456");
            u.setBirthday(formatter.parse("19901009"));
            u.setEmail("zhangsan@qq.com");
            u.setFriends(friends);
```

```
User f1 = new User();
f1.setName("李四");
f1.setPassword("123456");
f1.setBirthday(formatter.parse("19891009"));
f1.setEmail("lisi@qq.com");
User f2 = new User();
f2.setName("王五");
f2.setPassword("123456");
f2.setBirthday(formatter.parse("19881009"));
f2.setEmail("wangwu@qq.com");

/**
* JSON序列化
* java对象转JSON
*/
ObjectMapper mapper = new ObjectMapper();
String jsonObject = mapper.writeValueAsString(u);
System.out.println("---------------------------------------------");
System.out.println("java对象转JSON");
System.out.println(jsonObject);
System.out.println("---------------------------------------------");
/**
 * JSON序列化
 * java集合转JSON
 */
List<User> users = new ArrayList<User>();
users.add(u);
users.add(f1);
users.add(f2);
System.out.println("---------------------------------------------");
System.out.println("java集合转JSON");
String jsonlist = mapper.writeValueAsString(users);
System.out.println(jsonlist);
System.out.println("---------------------------------------------");
/**
 * JSON序列化
 * java数组对象混合转JSON
 * u有好友f1,f2
 * friends.add(f1); friends.add(f2);
 */
friends.add(f1);
friends.add(f2);
String json = mapper.writeValueAsString(u);
System.out.println("---------------------------------------------");
System.out.println("java数组对象混合转JSON");
System.out.println(json);
System.out.println("---------------------------------------------");

/**
 * JSON反序列化
 * ObjectMapper支持byte[]、File、InputStream、字符串等数据的JSON反序列化
```

```
        */
        String jsonString = "{\"name\": \"张三\",\"password\": \"123456\",\"birthday
\": 655401600000,\"email\": \"zhangsan@qq.com\",\"friends\": [     ]}";
        ObjectMapper mapper2 = new ObjectMapper();
        User user = mapper2.readValue(jsonString, User.class);
        System.out.println(user.toString());
    }
}
```

运行结果：

Java 对象转 JSON：

```
{
    "name": "张三",
    "password": "123456",
    "birthday": 655401600000,
    "email": "zhangsan@qq.com",
    "friends": []
}
```

Java 集合转 JSON：

```
[
    {
        "name": "张三",
        "password": "123456",
        "birthday": 655401600000,
        "email": "zhangsan@qq.com",
        "friends": []
    },
    {
        "name": "李四",
        "password": "123456",
        "birthday": 623865600000,
        "email": "lisi@qq.com",
        "friends": null
    },
    {
        "name": "王五",
        "password": "123456",
        "birthday": 592329600000,
        "email": "wangwu@qq.com",
        "friends": null
    }
]
```

Java 数组对象混合转 JSON：

```
{
    "name": "张三",
    "password": "123456",
    "birthday": 655401600000,
    "email": "zhangsan@qq.com",
    "friends": [
```

```
        {
            "name": "李四",
            "password": "123456",
            "birthday": 623865600000,
            "email": "lisi@qq.com",
            "friends": null
        },
        {
            "name": "王五",
            "password": "123456",
            "birthday": 592329600000,
            "email": "wangwu@qq.com",
            "friends": null
        }
    ]
}
```

JSON 反序列化：

```
姓名：张三
密码：123456
出生日期：Tue Oct 09 00:00:00 CST 1990
邮箱：zhangsan@qq.com
朋友[]
```

- ObjectMapper是JSON操作的核心，Jackson的所有JSON操作都是在ObjectMapper中实现的。
- ObjectMapper有多个JSON序列化的方法，可以把JSON字符串保存在File、OutputStream等不同的介质中。
- writeValue(File arg0, Object arg1)把arg1转成json序列，并保存到arg0文件中。
- writeValue(OutputStream arg0, Object arg1)把arg1转成JSON序列，并保存到arg0输出流中。
- writeValueAsBytes(Object arg0)把arg0转成JSON序列，并把结果输出成字节数组。
- writeValueAsString(Object arg0)把arg0转成JSON序列，并把结果输出成字符串。

5.6 课后小结

1. 对序列化的理解

- 通过序列化和反序列化实现了对象状态的保存、传输以及对象的重建。在进行对象序列化时，开发人员可以根据自身情况，灵活选择默认方式或者自定义方式实现对象的序列化和反序列化。
- 序列化机制是 Java 中对轻量级持久化的支持。
- 序列化的字节流数据在网上传输的安全问题需要引起大家足够的注意。
- 序列化破坏了原有类的数据的“安全性”，例如 private 属性是不起作用的。

2. 序列化带来的问题

（1）网络传输的安全性。

对象进行序列化之后转化成有序的字节流在网络上进行传输，如果通过默认的序列化方式，则代码都是以明文的方式进行传输。这种情况下，部分字段的安全性是不能保障的，特别是像密码这样的安全敏感的信息。因此，如果您需要对部分字段信息进行特殊的处理，那么应当选择定制对象的序列化方式，例如对密码等敏感信息进行加密处理。

（2）类自身封装的安全性。

对对象进行序列化时，类中所定义的被 private、final 等访问控制符所修饰的字段是直接忽略这些访问控制符而直接进行序列化的，因此，原本在本地定义的想要一次控制字段的访问权限的工作都是不起作用的。对于序列化后的有序字节流来说一切都是可见的，而且是可重建的。这在一定程度上削弱了字段的安全性。因此，如果您需要特别处理这些信息，可以选择相应的方式对这些属性进行加密或者其他可行的处理，以尽量保持数据的安全性。

所以并非所有的对象都可以序列化。有很多原因，比如下列几点。

① 安全方面的原因，比如一个对象拥有 private、public 等 field，对于一个要传输的对象，比如写到文件，或者进行 rmi 传输等，在序列化进行传输的过程中，这个对象的 private 等域是不受保护的。

② 资源分配方面的原因，比如 socket、thread 类，如果可以序列化，进行传输或者保存，也无法对它们进行重新的资源分配，而且也没有必要这样实现。

3. 序列化注意事项

关于对象的序列化，总结下注意事项，如下所示。

- 对象的类名、Field（包括基本类型、数组及对其他对象的引用）都会被序列化，对象的 static Field、transient Field 及方法不会被序列化；
- 实现 Serializable 接口的类，如不想某个 Field 被序列化，可以使用 transient 关键字进行修饰；
- 保证序列化对象的引用类型 Filed 的类也是可序列化的，如不可序列化，可以使用 transient 关键字进行修饰，否则会序列化失败；
- 反序列化时必须要有序列化对象的类的 class 文件；
- 当通过文件网络读取序列化对象的时候，必须按写入的顺序来读取。

5.7 课后习题

一、填空题

1. 若要用 ObjectOutputStream 写入一个对象，那么这个对象必须实现________接口，不然程序会抛出 NoSerializableException 类型的异常。

2. 对象的输出流将指定的对象写入到文件的过程，就是将对象________的过程，对象的输入流将指定序列化好的文件读出来的过程，就是对象________的过程。

3. ObjectOutputStream 类扩展 DataOutput 接口。writeObject() 方法是最重要的方法，用于________。

二、选择题

1. 以下关于对象序列化描述正确的是________。

A. 使用 FileOutputStream 可以将对象进行传输
B. 使用 PrintWriter 可以将对象进行传输
C. 使用 ObjectOutputStream 类完成对象存储，使用 ObjectInputStream 类完成对象读取
D. 对象序列化的所属类需要实现 Serializable 接口
2. 以下关于序列化的描述正确的是________。
A. 所有 Java 对象都可序列化
B. 所有 Java 对象都必须序列化
C. Java 对象的所有成员和方法都要序列化
D. Java 对象是根据需要时才序列化
3. 序列化时使用 FileOutputStream 对象的________方法。
A. writeObject()　　B. readObject()
C. write()　　D. read()
4. 可序列化类中的 writeObject() 和 readObject() 方法使用________访问控制修饰符。
A. public　　B. 没有访问控制修饰符
C. protected　　D. private

三、简答题

1. 什么是 Java 序列化，如何实现 Java 序列化？
2. 在什么情况下需要序列化？
3. Java 中实现序列化有哪几种方法？

5.8 上机实训

实训一 序列化的使用。

1. 编写一个可以序列化的个人账户类 AccountRecordSerializable，其具有如下的成员变量：

```
privateint account;
private String firstName;
private String lastName;
private double balance;
```

具有如下的成员方法：

```
AccountRecordSerializable(int acct, String first, String last, double bal) //初始化
 int getAccount()
double getBalance()
String  getFirstName()
String  getLastName()
void setAccount(int acct)
void setBalance(double bal)
void setFirstName(String first)
void setLastName(String last)
```

2. 编写一个类 CreateSequentialFile，将上述的 AccountRecordSerializable 对象写入到顺序文件中。成员方法为

```
    void  openFile() //使用Fileoutputstream和Objectoutputstream创建和打开磁盘文件clients.ser
    void   addRecords()  //用户通过键盘依次输入account  firstnamelasname  balance构建
AccountRecordSerializable类对象，然后序列化到文件中。可以输入多行信息。用户按下[Ctrl+Z]组合键结束输入
     void  closeFile() //关闭磁盘文件对象
```

3. 编写测试类 TestCreateSequentialFile

main 方法中创建 CreateSequentialFile 类对象，然后调用 openFile()、addRecords()、closeFile() 方法。

运行结果如图 5-6 所示。

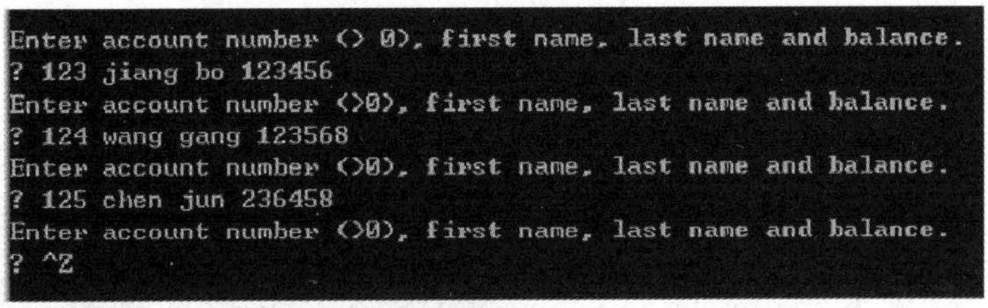

```
Enter account number (> 0), first name, last name and balance.
? 123 jiang bo 123456
Enter account number (>0), first name, last name and balance.
? 124 wang gang 123568
Enter account number (>0), first name, last name and balance.
? 125 chen jun 236458
Enter account number (>0), first name, last name and balance.
? ^Z
```

图5-6　运行结果

实训二　编写程序，实现学生信息及成绩的读取和写入，要求用对象序列化机制实现。

任务需求：

1. 学生信息及成绩保存到本地磁盘的“save.txt”中。
2. 学生的身份证号禁止序列化。
3. 运行时如果不存在“save.txt”则重新录入学生的信息与成绩，如图 5-7 所示。
4. 如果存在“save.txt”，显示学生信息、成绩和总成绩，如图 5-8 所示。

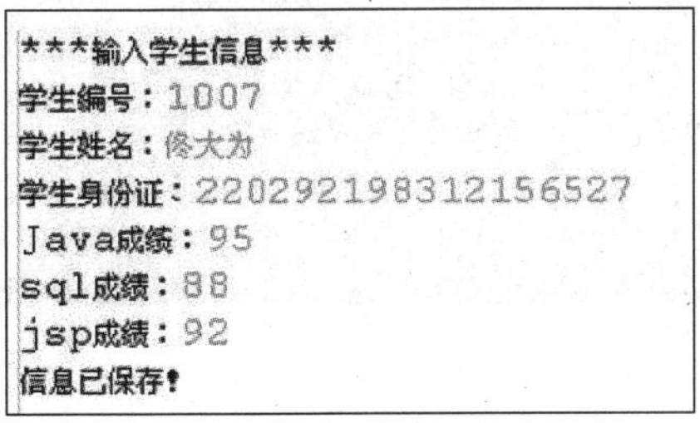

```
***输入学生信息***
学生编号：1007
学生姓名：佟大为
学生身份证：220292198312156527
Java成绩：95
sql成绩：88
jsp成绩：92
信息已保存!
```

图5-7　录入

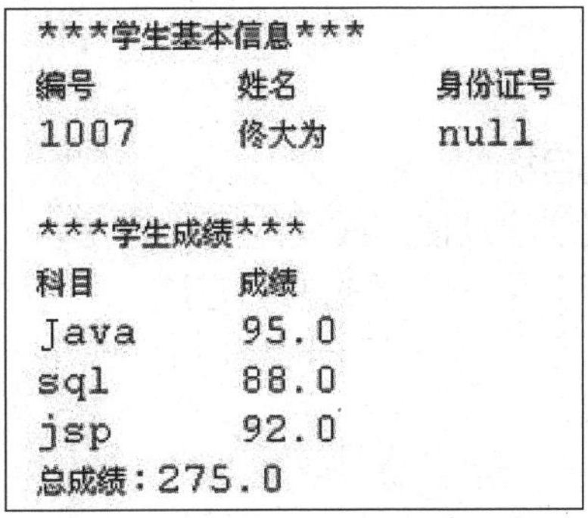

```
***学生基本信息***
编号      姓名      身份证号
1007      佟大为    null

***学生成绩***
科目      成绩
Java      95.0
sql       88.0
jsp       92.0
总成绩：275.0
```

图5-8　显示

5. 使用对象序列化机制实现。

实现思路：

1. 编写学生类、成绩类、科目类
2. 科目可以采用 Map 存储来初始化

实训三　请编写程序，把任意对象序列化成 JSON，如果属性不是基本数据类型，则需要序列化相关联的属性。并将 JSON 对象反序列化。

PART 06

知识领域6

Java多线程机制

知识目标

1. 理解进程、线程及多线程的概念。
2. 熟悉线程的5种状态及状态之间的转换关系。
3. 掌握线程的创建和启动方法。
4. 掌握线程的优先级设置及线程的常用调度方法。
5. 理解线程的同步机制。

■ 能力目标

1. 掌握创建及使用线程的两种方法。
2. 熟练使用线程类的常用方法。
3. 熟练使用线程的同步机制完成相应的编码。

■ 素质目标

1. 能够阅读科技文档和撰写分析文档。
2. 能够查阅JDK API。
3. 增强学生团队协作能力。

6.1 应用场景

计算机可以同时完成多项任务，称为并发。并发完成的每个任务就是一个独立线程。在网络分布式、高并发应用程序的情况下，Java 多线程编程技术在很多开发工作中得到了非常广泛的应用。下面列举几个经典的多线程问题。

1. 火车票预订问题

假定火车票有 10 000 张，现有 10 个售票点。每个售票窗口就像一个线程，它们各自运行，共同访问相同的数据——火车票的数量。由于多个线程并发地执行，访问共享同一数据，会出现数据不一致的现象，所有必须要用同步锁 synchronized，保证某一时刻只能有一个线程执行售票功能。

2. 经典生产者与消费者问题

生产者不断地往仓库中存放产品，消费者从仓库中消费产品。其中生产者和消费者都可以有若干个。

在这里，生产者是一个线程，消费者是一个线程。仓库容量有限，只有库满时生产者不能存放，库空时消费者不能取产品，这就是线程的同步。

3. 客户机 / 服务器通信问题

假如 Server 只能接受一个 Client 请求，当第一个 Client 连接后就占据了这个位置，后续 Client 不能再继续连接。采用多线程机制，当 Server 每接受到一个 Client 连接请求之后，都把处理流程放到一个独立的线程里去运行，然后等待下一个 Client 连接请求，这样就不会阻塞 Server 端接收请求了。

4. 多线程死锁问题

当一个线程永远地持有一个锁，并且其他线程都尝试去获得这个锁时，那么它们将永远被阻塞，这个我们都知道。如果线程 A 持有锁 L 并且想获得锁 M，线程 B 持有锁 M 并且想获得锁 L，那么这两个线程将永远等待下去，这种情况就是最简单的死锁形式。

5. 异步处理问题

假设有个请求，这个请求服务端的处理需要执行 3 个很缓慢的 IO 操作（比如数据库查询或文件查询），那么正常的顺序可能如下所示（括号里面代表执行时间）。

① 读取文件 1　　（10ms）
② 处理 1 的数据　　（1ms）
③ 读取文件 2　　（10ms）
④ 处理 2 的数据　　（1ms）
⑤ 读取文件 3　　（10ms）
⑥ 处理 3 的数据　　（1ms）
⑦ 整合 1、2、3 的数据结果　　（1ms）

如果采用单个线程只能串行处理，单线程总共就需要 34ms。如果采用多线程，①②、③④、⑤⑥分别分给 3 个线程去做，就只需要 12ms 了。

这里把一个项目分解成多个任务，每个任务分配一个线程，就是异步处理机制。

6.2 相关知识

6.2.1 相关概念

进程是指运行中的应用程序，每个进程都有自己独立的内存空间。一个应用程序可以同时启动多个线程。

几乎所有的操作系统都支持同时运行多个任务，一个任务通常就是一个程序，每个运行中的程序就是一个进程。当一个程序运行时，内部可能包含了多个顺序执行流，每个顺序执行流就是一个线程。一个进程可以由多个线程组成，即在一个进程中可以同时运行多个不同的线程，它们分别执行不同的任务。当进程内的多个线程同时运行时，这种运行方式被称为并发运行。

1. 进程（Process）

进程是计算机中的程序关于某数据集合上的一次运行活动，是系统进行资源分配和调度的基本单位，是操作系统结构的基础。在早期面向进程设计的计算机结构中，进程是程序的基本执行实体；在当代面向线程设计的计算机结构中，进程是线程的容器。程序是指令、数据及其组织形式的描述，进程是程序的实体。

2. 进程的特征

动态性：进程的实质是程序在多道程序系统中的一次执行过程，进程是动态产生、动态消亡的。

并发性：任何进程都可以同其他进程一起并发执行。

独立性：进程是一个能独立运行的基本单位，同时也是系统分配资源和调度的独立单位。

异步性：由于进程间的相互制约，使进程具有执行的间断性，即进程按各自独立的、不可预知的速度向前推进。

结构特征：进程由程序、数据和进程控制块 3 部分组成。

多个不同的进程可以包含相同的程序：一个程序在不同的数据集里就构成不同的进程，能得到不同的结果；但是执行过程中，程序不能发生改变。

3. 线程（Thread）

线程有时被称为轻量级进程（Lightweight Process，LWP），是程序执行流的最小单元。一个标准的线程由线程 ID、当前指令指针（PC）、寄存器集合和堆栈组成。另外，线程是进程中的一个实体，是被系统独立调度和分派的基本单位，线程自己不拥有系统资源，只拥有一点在运行中必不可少的资源，但它可与同属一个进程的其他线程共享进程所拥有的全部资源。一个线程可以创建和撤消另一个线程，同一进程中的多个线程之间可以并发执行。由于线程之间的相互制约，致使线程在运行中呈现出间断性。线程也有就绪、阻塞和运行 3 种基本状态。就绪状态是指线程具备运行的所有条件，逻辑上可以运行，在等待处理机；运行状态是指线程占有处理机正在运行；阻塞状态是指线程在等待一个事件（如某个信号量），逻辑上不可执行。每一个程序都至少有一个线程，若程序只有一个线程，那就是程序本身。

线程是程序中一个单一的顺序控制流程，是进程内一个相对独立的、可调度的执行单元，是系统独立调度和分派 CPU 的基本单位，也是运行中的程序的调度单位。在单个程序中同时运行多个线程完成不同的工作，称为多线程。

一个线程。

4. 多线程（multithreading）

多线程是指从软件或者硬件上实现多个线程并发执行的技术。具有多线程能力的计算机因有硬件支持而能够在同一时间执行多于一个线程，进而提升整体处理性能。具有这种能力的系统包括对称多处理机、多核心处理器以及芯片级多处理（Chip-level multithreading）或同时多线程（Simultaneous multithreading）处理器。在一个程序中，这些独立运行的程序片段叫作“线程”（Thread），利用它编程的概念就叫作“多线程处理（Multithreading）”。具有多线程能力的计算机因有硬件支持而能够在同一时间执行多于一个线程，进而提升整体处理性能。

在实际的应用中，多线程是非常有用的，一个浏览器必须能同时下载多个图片，即一个 Web 服务器必须能同时响应多个用户请求，Java 虚拟机本身就在后台提供一个超级线程来进行垃圾回收。总之，多线程在实际编程中的应用是非常广泛的。

6.2.2 线程的创建和启动

线程的创建有两种方式：通过继承 Thread 类来创建和通过实现 Runnable 接口来创建。

1. 继承 Thread 类创建线程类

通过继承 Thread 类创建线程类的具体步骤和具体代码如下。

- 定义一个继承 Thread 类的子类，并重写该类的 run() 方法；
- 创建 Thread 子类的实例，即创建了线程对象；
- 调用该线程对象的 start() 方法启动线程。

```
class MyThead extends Thread   {
    public void run()    {
     //do something here
    }
 }

public static void main(String[] args){
 MyThread oneThread = new MyThread();
   //启动线程
 oneThread.start();
}
```

例程分析：

Thread_1.java

```
package com.hbliti.thread;
public class Thread_1 extends Thread{
    //重写run方法，同时也是线程的执行体
    public void run(){
        for(int i = 0;i<1000;i++){
            //这里得到的是正在执行的线程的名字
            System.out.println(this.getName()+":"+i);
        }
    }
    public static void main(String[] args) {
        //通过无参构造函数创建实例
        Thread_1 t1 = new Thread_1();
```

```
        //设置线程的名字
        t1.setName("线程1");
        //通过调用start()方法来启动线程，线程启动后将执行run()方法体内的代码
        t1.start();
        for(int i =0;i<1000;i++){
            System.out.println(Thread.currentThread().getName()+":"+i);
        }
    }
}
```

运行结果如图 6-1 所示。在学习多线程之前，我们知道程序是顺序执行的，在我们调用 start 方法后应该先执行线程的执行体（即循环输出 1000 次），但我们可以看到主线程和我们创建的线程是交替执行的，这就体现了线程的并发性。

```
<terminated> Thread_1 [Java Application] C:\Program Files\Java\jdk1.7.0_17\bin\javaw.exe (2017
main:0
线程1:0
main:1
线程1:1
main:2
线程1:2
main:3
线程1:3
main:4
线程1:4
main:5
线程1:5
main:6
线程1:6
main:7
线程1:7
线程1:8
线程1:9
main:8
main:9
main:10
```

图6-1　多线程的创建与启动（使用继承Thread类）

2. 实现 Runnable 接口创建线程类

通过实现 Runnable 接口创建线程类的具体步骤和具体代码如下。

- 定义 Runnable 接口的实现类，并重写该接口的 run() 方法。
- 创建 Runnable 实现类的实例，并以此实例作为 Thread 的 target 对象，即该 Thread 对象才是真正的线程对象。
- 调用该线程对象的 start() 方法启动线程。

```
class MyRunnable implements Runnable   {
  public void run()   {
  //do something here
  }
}
Runnable oneRunnable = new MyRunnable();
Thread oneThread = new Thread(oneRunnable);
oneThread.start();
```

例程分析：

Thread_2.java

```java
package com.hbliti.thread;
//实现Runnable接口
public class Thread_2 implements Runnable{
    public void run() {
        for(int i = 0;i<1000;i++){
            System.out.println(Thread.currentThread().getName()+":"+i);
        }
    }
    public static void main(String[] args) {
        //实例Thread_2对象
        Thread_2 th2 = new Thread_2();
        //通过带参的构造函数实例Thread对象，并指定线程名称
        Thread t2 = new Thread(th2,"线程2");
        Thread t3 = new Thread(th2,"线程3");
        t2.start();
        t3.start();
    }
}
```

运行结果如图 6-2 所示。

```
<terminated> Thread_2 [Java Application] C:\Program Files\Java\jdk1.7.0_17\bin\javaw.exe (2017
线程2:0
线程3:0
线程2:1
线程3:1
线程2:2
线程3:2
线程2:3
线程3:3
线程2:4
线程3:4
线程2:5
线程3:5
线程2:6
线程3:6
线程2:7
线程3:7
线程2:8
线程3:8
线程2:9
线程3:9
线程2:10
```

图6-2　多线程的创建与启动（使用实现Ruanable接口）

3. 继承 Thread 类和实现 Runnable 接口的区别

实现 Runnable 接口的方式如下。

- 线程实现了 Runnable 接口后还可以继承其他的类。
- 多个线程之间共享一个实例，这种情况适合多个线程来处理同一份资源。

使用继承 Thread 的方式如下。

- 弊端：由于继承了 Thread 类，就不能再继承其他类。
- 优点：编程简单，要访问当前线程，直接用 this 即可。

在实际应用中多数情况都采用实现 Runnable 接口的方式。

6.2.3 线程的生命周期

线程在它的生命周期中会处于不同的状态。当线程被创建后，它要经过新建、就绪（可运行）、运行、阻塞和死亡 5 种状态。尤其是当线程启动后，它不能一直处于运行状态，所以线程要在多条线程之间进行转换，于是线程会在多次运行、阻塞之间进行转换。状态的转换如图 6-3 所示。

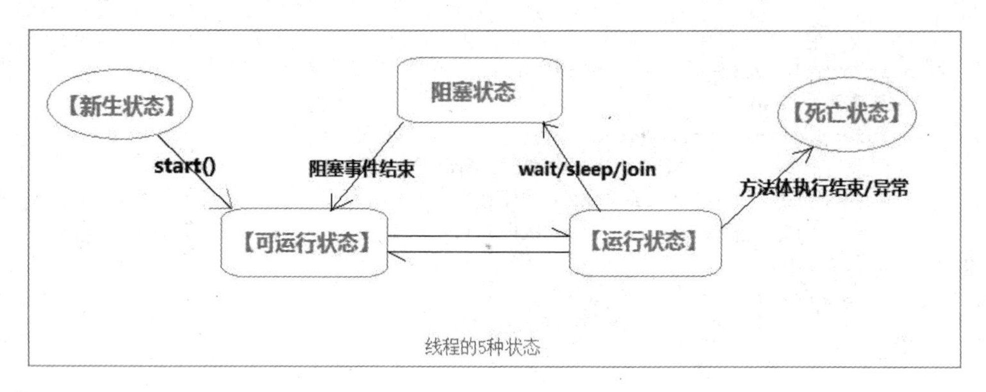

图6-3　线程的状态图

1. 新建状态

用 new 关键字和 Thread 类或其子类建立一个线程对象后，该线程对象就处于新生状态。处于新生状态的线程有自己的内存空间，通过调用 start 方法进入就绪状态（runnable）。

不能对已经启动的线程再次调用start()方法，否则会出现Java.lang.IllegalThreadStateException异常。

2. 就绪状态

处于就绪状态的线程已经具备了运行条件，但还没有分配到 CPU，处于线程就绪队列（尽管是采用队列形式，事实上，把它称为可运行池而不是可运行队列。因为 CPU 的调度不一定是按照先进先出的顺序来调度的），等待系统为其分配 CPU。等待状态并不是执行状态，当系统选定一个等待执行的 Thread 对象后，它就会从等待执行状态进入执行状态，系统挑选的动作称之为“CPU 调度”。一旦获得 CPU，线程就进入运行状态并自动调用自己的 run 方法。

如果希望子线程调用start()方法后立即执行，可以使用Thread.sleep()方式使主线程睡眠一会儿，转去执行子线程。

3. 运行状态

处于运行状态的线程最为复杂，它可以变为阻塞状态、就绪状态和死亡状态。处于就绪状态的线程，如果获得了 CPU 的调度，就会从就绪状态变为运行状态，执行 run() 方法中的任务。如果该线程失去了 CPU 资源，就会又从运行状态变为就绪状态。重新等待系统分配资源。也可以对在运行状态的线程调用 yield() 方法，它就会让出 CPU 资源，再次变为就绪状态。

注意：当发生如下情况时，线程会从运行状态变为阻塞状态。

① 线程调用 sleep 方法主动放弃所占用的系统资源。

② 线程调用一个阻塞式 IO 方法，在该方法返回之前，该线程被阻塞。

③ 线程试图获得一个同步监视器，但该同步监视器正被其他线程所持有。

④ 线程在等待某个通知（notify）。

⑤ 程序调用了线程的 suspend 方法将线程挂起。不过该方法容易导致死锁，所以程序应该尽量避免使用该方法。

当线程的 run() 方法执行完，或者被强制性地终止，例如出现异常，或者调用了 stop()、desyory() 方法等，就会从运行状态转变为死亡状态。

4. 阻塞状态

处于运行状态的线程在某些情况下，如执行了 sleep（睡眠）方法，或等待 I/O 设备等资源，将让出 CPU 并暂时停止自己的运行，进入阻塞状态。

在阻塞状态的线程不能进入就绪队列。只有当引起阻塞的原因消除时，如睡眠时间已到，或等待的 I/O 设备空闲下来，线程便转入就绪状态，重新到就绪队列中排队等待，被系统选中后从原来停止的位置开始继续运行。

5. 死亡状态

当线程的 run() 方法执行完，或者被强制性地终止，就认为它死去。这个线程对象也许是活的，但是，它已经不是一个单独执行的线程。线程一旦死亡，就不能复生。如果在一个死去的线程上调用 start() 方法，会抛出 Java.lang.IllegalThreadStateException 异常。

例程分析：

ThreadState.java

```
package com.hbliti.thread;
public class ThreadState implements Runnable {
    public synchronized void waitForASecond() throws InterruptedException {
        wait(500); // 使当前线程等待0.5秒或其他线程调用notify()或notifyAll()方法
    }

    public synchronized void waitForYears() throws InterruptedException {
        wait(); // 使当前线程永久等待，直到其他线程调用notify()或notifyAll()方法
    }

    public synchronized void notifyNow() throws InterruptedException {
        notify(); // 唤醒由调用wait()方法进入等待状态的线程
    }

    public void run() {
        try {
            waitForASecond(); // 在新线程中运行waitForASecond()方法
            waitForYears(); // 在新线程中运行waitForYears()方法
        } catch (InterruptedException e) {
            e.printStackTrace();
        }
    }
}
```

Test.java

```java
package com.hbliti.thread;
public class Test {
    public static void main(String[] args) throws InterruptedException {
        ThreadState state = new ThreadState();// 创建State对象
        Thread thread = new Thread(state);// 利用State对象创建Thread对象
        System.out.println("新建线程: " + thread.getState());// 输出线程状态
        thread.start(); // 调用thread对象的start()方法，启动新线程
        System.out.println("启动线程: " + thread.getState());// 输出线程状态
        Thread.sleep(100); // 当前线程休眠0.1秒，使新线程运行waitForASecond()方法
        System.out.println("计时等待: " + thread.getState());// 输出线程状态
        Thread.sleep(1000); // 当前线程休眠1秒，使新线程运行waitForYears()方法
        System.out.println("等待线程: " + thread.getState());// 输出线程状态
        state.notifyNow(); // 调用state的notifyNow()方法
        System.out.println("唤醒线程: " + thread.getState());// 输出线程状态
        Thread.sleep(1000); // 当前线程休眠1秒，使新线程结束
        System.out.println("终止线程: " + thread.getState());// 输出线程状态
    }
}
```

运行结果如图 6-4 所示。

```
<terminated> Test (3) [Java Application] C:\Program Files\Java\jdk1.7.0_17\bin\javaw.exe (2017-
新建线程: NEW
启动线程: RUNNABLE
计时等待: TIMED_WAITING
等待线程: WAITING
唤醒线程: BLOCKED
终止线程: TERMINATED
```

图6-4 线程的状态转换运行结果图

6.2.4 线程的管理

1. join 线程

Thread 类提供了让一个线程等待另一个线程完成的方法——join() 方法。当某个程序执行流中调用了其他线程的 join() 方法的时候，调用线程将被阻塞，直到 join 方法加入的 join 线程执行完成为止。

例程分析：

TestJoin.java

```java
package com.hbliti.thread;
public class TestJoin extends Thread{
    public void run(){
        for(int i = 0;i<10;i++){
            System.out.println(currentThread().getName()+":"+i);
        }
    }
```

```
    public static void main(String[] args) throws InterruptedException {
        for(int i = 0;i<30;i++){
            System.out.println(currentThread().getName()+":"+i);
            if(i==10){
                TestJoin tj = new TestJoin();
                //启动要加入的线程
                tj.start();
                //调用join()方法
                tj.join();
            }
        }
    }
}
```

执行结果如图 6-5 所示。

```
<terminated> TestJoin [Java Application] C:\Program Files\Java\jdk1.7.0_17\bin\javaw.exe (2017-
main:0
main:1
main:2
main:3
main:4
main:5
main:6
main:7
main:8
main:9
main:10
Thread-0:0
Thread-0:1
Thread-0:2
Thread-0:3
Thread-0:4
Thread-0:5
Thread-0:6
Thread-0:7
Thread-0:8
Thread-0:9
main:11
main:12
main:13
main:14
main:15
```

图6-5　运行结果

我们可以看到当条件满足，调用 join() 方法后，要等到执行完加入的线程的线程体，再继续 main 线程。

2. 线程睡眠

当我们要让线程暂停一段时间时，我们可以让线程睡眠，进入阻塞状态。方法为：

```
static void sleep(long millis)//让当前正在执行的线程暂停millis秒，并进入阻塞状态
```

实例：每隔一秒输出一次。

TestSleep.java

```
package com.hbliti.thread;
public class TestSleep{
```

```
    public static void main(String[] args) throws InterruptedException {
        for(int i = 10;i>0;i--){
            System.out.println("倒计时: "+i);
            Thread.sleep(1000);
        }
    }
}
```

效果为每过一秒输出一次。

3. 改变线程优先级

每个线程执行时都有一定的优先级，优先级高的获得较多的执行机会，优先级低的线程获得较少的执行机会。

每个线程默认的优先级都与创建它的父线程具有相同的优先级，在默认情况下，main 的优先级是 5，由 main 创建的线程的优先级也是 5。

Thread 类提供了 setPriority(int i) 和 getPriority(int i) 方法来设置和获得优先级。其中 setPriority 方法的参数可以是一个整数，也可以是 Thread 类的 3 个静态常量。常量包括：

MAX_PRIORITY：其值是 10。

MIN_PRIORITY：其值是 1。

NORM_PRIORITY：其值是 5。

例程分析：

TestPriority.java

```
package com.hbliti.thread;
public class TestPriority extends Thread{

    public void run(){
        for(int i = 0;i<1000;i++){
            System.out.println(currentThread().getName()+":"+i);
        }
    }
    public static void main(String[] args) {
        System.out.println(currentThread().getName()+"的优先级是:"+currentThread().getPriority());
        TestPriority tp1 = new TestPriority();
        System.out.println(tp1.getName()+"的优先级是:"+tp1.getPriority());
        //改变main线程的优先级
        currentThread().setPriority(MIN_PRIORITY);
        System.out.println(currentThread().getName()+"的优先级是:"+currentThread().getPriority());
        TestPriority tp2 = new TestPriority();
        System.out.println(tp2.getName()+"的优先级是: "+tp2.getPriority());
        tp1.start();
        tp2.start();
    }
}
```

运行结果如图 6-6 所示。

```
<terminated> TestPriority [Java Application] C:\Program Files\Java\jdk1.7.0_17\bin\javaw.exe (2017
main的优先级是：5
Thread-0的优先级是：5
main的优先级是：1
Thread-1的优先级是：1
Thread-0:0
Thread-0:1
Thread-0:2
Thread-0:3
Thread-0:4
Thread-0:5
Thread-0:6
Thread-1:0
Thread-0:7
Thread-1:1
Thread-0:8
Thread-1:2
Thread-0:9
Thread-1:3
Thread-0:10
Thread-1:4
Thread-0:11
Thread-1:5
Thread-0:12
Thread-1:6
Thread-0:13
Thread-1:7
Thread-0:14
Thread-1:8
Thread-0:15
Thread-1:9
Thread-0:16
Thread-1:10
Thread-0:17
Thread-1:11
Thread-0:18
Thread-1:12
Thread-0:19
```

图6-6　运行结果

6.3　任务实施

任务　Java 多线程并发控制——模拟火车票售票

当多个线程同时操作一个可共享的资源变量时（如数据的增删改查），将会导致数据不准确，相互之间产生冲突，因此 Java 多线程并发控制加入同步锁以避免在该线程没有完成操作之前被其他的线程调用，从而保证了该变量的唯一性和准确性。

1. 任务需求

模拟火车站售票大厅销售火车票。

2. 任务分析

票数要使用同一个静态值；为保证不会出现卖出同一张票，要 Java 多线程同步锁。

（1）创建一个售票窗口类 TicketWindow，实现 Runable 接口，重写 run 方法，在 run 方法里面执行售票操作。售票要使用同步锁，即有一个窗口卖这张票时，其他窗口要等这张票卖完。

（2）创建售票大厅类 TestTicketLobby，主方法创建 4 个售票窗口，创建线程，并启动线程，开设售票。

其类图如图 6-7 所示。

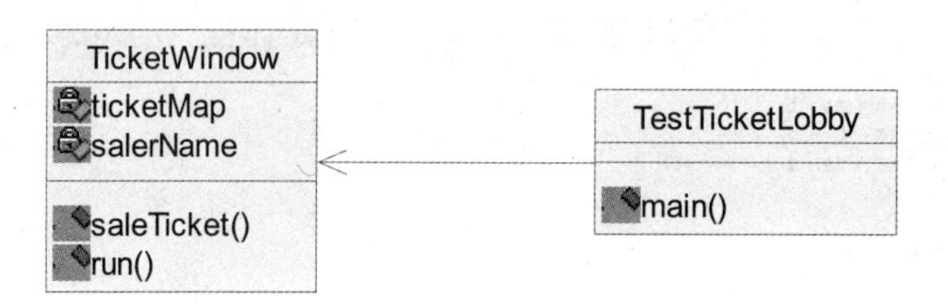

图6-7　类图

3. 任务实现

售票窗口类 TicketWindow。

```
package com.daiinfo.seniorjava.ken6.implement;
import java.util.Iterator;
import java.util.Map;
/**
 * 售票窗口类，每个窗口包含一个售票员和票池
 *
 * @author 戴远泉
 *
 */
1public class TicketWindow implements Runnable {
    private Map<String, Boolean> ticketMap;// 票池
    private String salerName;// 售票员姓名
    public TicketWindow(Map<String, Boolean> ticketMap, String salerName) {
        this.ticketMap = ticketMap;
        this.salerName = salerName;
    }
    // 售票
    private void saleTicket() {
        for (Iterator<String> it = ticketMap.keySet().iterator();;) {
            synchronized (ticketMap) {
                if (it.hasNext()) {
                    String ticketNo = it.next();
                    if (!ticketMap.get(ticketNo)) {
                        System.out
                                .println(salerName + ":" + ticketNo + "已被售出。");
                        ticketMap.put(ticketNo, true);
                        try {
                           //让当前线程睡眠片刻，使得别的线程有机会执行
                           Thread.sleep(100);
                        } catch (InterruptedException e) {
                            // TODO Auto-generated catch block
                            e.printStackTrace();
                        }
                    }
                } else {
                    break;
                }
            }
        }
```

```
    }
    @Override
    public void run() {
        saleTicket();
    }
}
```

售票大厅类 TestTicketLobby。

```
package com.daiinfo.seniorjava.ken6.implement;
import java.util.HashMap;
import java.util.Map;
import java.util.concurrent.ExecutorService;
import java.util.concurrent.Executors;
/**
 * 售票大厅，开设多个售票窗口，每个窗口设一个售票员。每个窗口是一个线程，独立售票
 *
 * @author 戴远泉
 *
 */
public class TestTicketLobby {
    public static void main(String[] args) {
        // 票池：<票编号，是否已出售>
        Map ticketMap = new HashMap<String, Boolean>();
        // 生成1000张火车票到票池
        for (int i = 1; i <= 1000; i++) {
            ticketMap.put("T" + i, false);
        }
        // 生成4个售票窗口
        TicketWindow s1 = new TicketWindow(ticketMap, "张三");
        TicketWindow s2 = new TicketWindow(ticketMap, "李四");
        TicketWindow s3 = new TicketWindow(ticketMap, "王五");
        TicketWindow s4 = new TicketWindow(ticketMap, "何六");
        // 每个窗口创建一个线程
        Thread t1 = new Thread(s1);
        Thread t2 = new Thread(s2);
        Thread t3 = new Thread(s3);
        Thread t4 = new Thread(s4);
        // java通过Executors提供一个可缓存线程池
        ExecutorService service = Executors.newCachedThreadPool();
        service.execute(t1);
        service.execute(t2);
        service.execute(t3);
        service.execute(t4);
        // 执行完线程池中的线程后尽快退出
        service.shutdown();
    }
}
```

运行结果如图 6-8 所示。

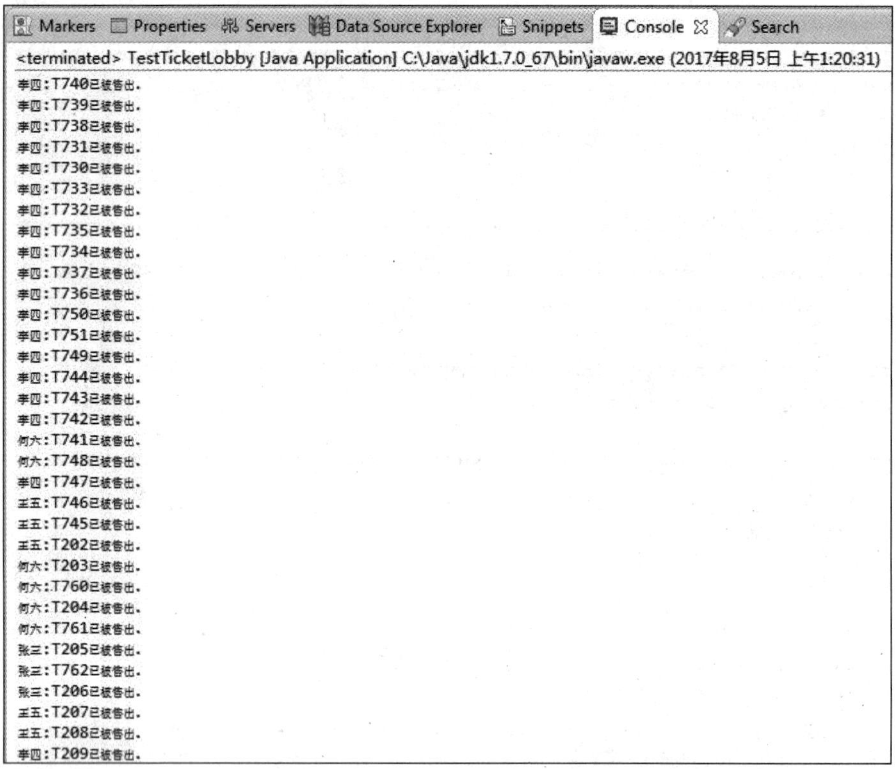

```
Markers  Properties  Servers  Data Source Explorer  Snippets  Console  Search
<terminated> TestTicketLobby [Java Application] C:\Java\jdk1.7.0_67\bin\javaw.exe (2017年8月5日 上午1:20:31)
李四:T740已被售出.
李四:T739已被售出.
李四:T738已被售出.
李四:T731已被售出.
李四:T730已被售出.
李四:T733已被售出.
李四:T732已被售出.
李四:T735已被售出.
李四:T734已被售出.
李四:T737已被售出.
李四:T736已被售出.
李四:T750已被售出.
李四:T751已被售出.
李四:T749已被售出.
李四:T744已被售出.
李四:T743已被售出.
李四:T742已被售出.
何六:T741已被售出.
何六:T748已被售出.
李四:T747已被售出.
王五:T746已被售出.
王五:T745已被售出.
王五:T202已被售出.
何六:T203已被售出.
何六:T760已被售出.
何六:T204已被售出.
何六:T761已被售出.
张三:T205已被售出.
张三:T762已被售出.
张三:T206已被售出.
王五:T207已被售出.
王五:T208已被售出.
李四:T209已被售出.
```

图6-8　火车站售票运行结果

6.4　拓展知识

本节补充两个知识点。

1. wait 和 notify 方法

Wait、notify、notifyAll 这 3 个方法是 Object 类定义的。

- this.wait()：让当前线程进入休眠状态，直到被其他线程调用 notify/notifyAll 唤醒。
- this.notify()/notifyAll()：唤醒由当前对象调用 wait 进入休眠状态的线程。
- notify()：随机唤醒一个线程，notifyAll() 唤醒所有。

2. synchronized 关键字

锁对象：在线程体中，如果由于使用的资源被其他线程访问会导致数据错误，那么我们在只用这个资源之前，可以通过 synchronized 关键字将这个共享对象锁住，使用完成之后进行释放。

锁方法：使用 synchronized 关键字修饰一个方法，表示对这个方法实现同步，即不允许多个线程同时调用这个方法。实际上方法被同步指的是调用当前方法的对象被锁住。

如果使用 synchronized 锁方法，那么调用这些方法的对象必须是同一个对象。

6.5　拓展训练

1. 任务需求

用 Java 模拟生产者 - 消费者问题。生产者生产产品，存放在仓库里，消费者从仓库里消费产品。

2. 任务分析

（1）生产者仅仅在仓储未满时候生产，仓满则停止生产。

（2）消费者仅仅在仓储有产品时候才能消费，仓空则等待。

（3）当消费者发现仓储没产品可消费时候会通知生产者生产。

（4）生产者在生产出可消费产品的时候，应该通知等待的消费者去消费。

类图如图 6-9 所示。

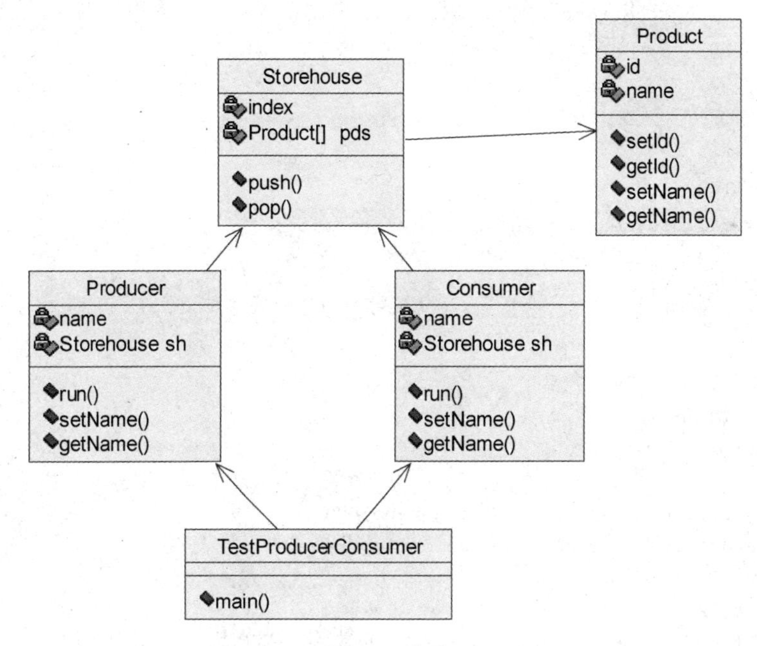

图6-9 类图

其中，TestProducerConsumer 是主类，Producer 为生产者，Consumer 为消费者，Product 为产品，Storehouse 为仓库。

3. 任务实现

Product.java

```
package com.daiinfo.seniorjava.ken6.prolongation;
public class Product {
    private int id;// 产品id
    private String name;// 产品名称
    public Product(int id) {
        this.id = id;
    }
    public Product(int id, String name) {
        this.id = id;
        this.name = name;
    }
    public int getId() {
        return id;
    }
    public void setId(int id) {
        this.id = id;
```

```
    }
    public String getName() {
        return name;
    }
    public void setName(String name) {
        this.name = name;
    }
    @Override
    public String toString() {
        return " 产品名称: " + name + "产品ID: " + id;
    }
}
```

Storehouse.java

```
package com.daiinfo.seniorjava.ken6.prolongation;
/**
 * 仓库类，用于存放产品
 * 采用栈结构
 * push，入栈，放产品，即生产产品
 * pop，出栈，取产品，即消费产品
 * @author  戴远泉
 * @version 1.0
 */
import java.util.concurrent.BlockingQueue;
import java.util.concurrent.LinkedBlockingQueue;
public class Storehouse {
    int index = 0;
    Product[] pds = new Product[10];// 构造数组，存放产品，容量是10
    // 放入仓库中，相当于入栈
    public synchronized void push(Product pd) {
        while (index == pds.length) {// 仓库满了，即栈满
            try {
                this.wait();// 让当前线程等待
            } catch (InterruptedException e) {
                // TODO Auto-generated catch block
                e.printStackTrace();
            }
        }
        pds[index] = pd;
        System.out.println(Thread.currentThread().getName() + " 生产了产品" + "["
                + pd.toString() + "]");
        this.index++;
        this.notify();// 唤醒在此对象监视器上等待的单个线程，即消费者线程
    }
    // 从仓库中拿出，相当于出栈
    public synchronized Product pop() {
        while (index == 0) {// 仓库空了，即栈空
            try {
                this.wait();
            } catch (InterruptedException e) {
                // TODO Auto-generated catch block
```

```
                e.printStackTrace();
            }
        }
        this.index--;// push第n个之后，this.index++，使栈顶为n+1，故return之前要减一
        System.out.println(Thread.currentThread().getName() + " 消费了产品" + "["
                + pds[index].toString() + "]");
        this.notify();
        return pds[index];
    }
}
```

Producer.java

```
package com.daiinfo.seniorjava.ken6.prolongation;
import java.util.Date;
public class Producer implements Runnable {
    private String name;
    private Storehouse sh = null;
    public Producer(String name, Storehouse sh) {
        this.name = name;
        this.sh = sh;
    }
    public String getName() {
        return name;
    }
    public void setName(String name) {
        this.name = name;
    }
    @Override
    public void run() {
        while (true) {
            Product product = new Product((int) (Math.random() * 10000)); // 产生0~9999之
间的随机整数
            sh.push(product);
            try {
                Thread.sleep(20);
            } catch (InterruptedException e1) {
                e1.printStackTrace();
            }
        }
    }
}
```

Consumer.java

```
package com.daiinfo.seniorjava.ken6.prolongation;
import java.util.Date;
public class Consumer implements Runnable {
    private String name;
    private Storehouse sh = null;
    public Consumer(String name, Storehouse sh) {
        this.name = name;
```

```
        this.sh = sh;
    }
    public String getName() {
        return name;
    }
    public void setName(String name) {
        this.name = name;
    }
    @Override
    public void run() {
        while (true) {
            Product product = sh.pop();
            try {
                Thread.sleep(2000);
            } catch (InterruptedException e) {
                e.printStackTrace();
            }
        }
    }
}
```

TestProducerConsumer.java

```
package com.daiinfo.seniorjava.ken6.prolongation;
import java.util.concurrent.ExecutorService;
import java.util.concurrent.Executors;
/**
 * java多线程模拟生产者消费者问题
 *
 * ProducerConsumer是主类，Producer是生产者，Consumer是消费者，Product是产品，Storehouse是仓库
 *
 * @author 戴远泉
 * @version 1.0 2017-8-7 上午02:49:02
 */
public class TestProducerConsumer {
    public static void main(String[] args) {
        Storehouse sh = new Storehouse();
        Producer p1 = new Producer("张三", sh);
        Producer p2 = new Producer("李四", sh);
        Consumer c1 = new Consumer("王五", sh);
        Consumer c2 = new Consumer("老刘", sh);
        Thread tp1 = new Thread(p1);// 新建一个生产者线程
        Thread tp2 = new Thread(p2);
        Thread tc1 = new Thread(c1);// 新建一个消费者线程
        Thread tc2 = new Thread(c2);
        /*
         * tp1.start();// 启动生产者线程
         *  tp2.start();
         *  tc1.start();// 启动消费者线程
         * tc2.start();
         */
```

```
        /**
         * 使用线程池
         */
        ExecutorService service = Executors.newCachedThreadPool();
        service.submit(tc1);
        service.submit(tc2);
        service.submit(tp1);
        service.submit(tp2);
    }
}
```

运行结果如图 6-10 所示。

```
Markers  Properties  Servers  Data Source Explorer  Snippets  Console  Search
<terminated> TestProducerConsumer [Java Application] C:\Java\jdk1.7.0_67\bin\javaw.exe (2017年8月15日 下午4:35:32)
pool-1-thread-3 生产了产品[ 产品名称: null产品ID: 5075]
pool-1-thread-1 消费了产品[ 产品名称: null产品ID: 5075]
pool-1-thread-4 生产了产品[ 产品名称: null产品ID: 192]
pool-1-thread-2 消费了产品[ 产品名称: null产品ID: 192]
pool-1-thread-4 生产了产品[ 产品名称: null产品ID: 4524]
pool-1-thread-3 生产了产品[ 产品名称: null产品ID: 8440]
pool-1-thread-4 生产了产品[ 产品名称: null产品ID: 5107]
pool-1-thread-3 生产了产品[ 产品名称: null产品ID: 186]
pool-1-thread-4 生产了产品[ 产品名称: null产品ID: 4355]
pool-1-thread-3 生产了产品[ 产品名称: null产品ID: 5223]
pool-1-thread-4 生产了产品[ 产品名称: null产品ID: 9608]
pool-1-thread-3 生产了产品[ 产品名称: null产品ID: 1528]
pool-1-thread-3 生产了产品[ 产品名称: null产品ID: 1452]
pool-1-thread-4 生产了产品[ 产品名称: null产品ID: 9132]
pool-1-thread-2 消费了产品[ 产品名称: null产品ID: 9132]
pool-1-thread-3 生产了产品[ 产品名称: null产品ID: 8893]
pool-1-thread-1 消费了产品[ 产品名称: null产品ID: 8893]
pool-1-thread-4 生产了产品[ 产品名称: null产品ID: 9718]
pool-1-thread-2 消费了产品[ 产品名称: null产品ID: 9718]
pool-1-thread-4 生产了产品[ 产品名称: null产品ID: 3452]
pool-1-thread-1 消费了产品[ 产品名称: null产品ID: 3452]
pool-1-thread-3 生产了产品[ 产品名称: null产品ID: 7739]
pool-1-thread-2 消费了产品[ 产品名称: null产品ID: 7739]
pool-1-thread-4 生产了产品[ 产品名称: null产品ID: 6563]
pool-1-thread-1 消费了产品[ 产品名称: null产品ID: 6563]
pool-1-thread-3 生产了产品[ 产品名称: null产品ID: 6802]
```

图6-10　运行结果

6.6　课后小结

1. 线程的创建

Java 中线程的创建主要有以下两种方式。

- 继承 Thread 类（Thread 中方法主要用于针对线程本身的处理）。
- 实现 Runnable 接口。

2. 线程的通信

线程之间的通信实际上是线程之间传递参数的问题。

- 往线程中传递参数：通过新建线程的构造函数。
- 线程执行中，往外界传递参数。
    - ✓ 通过方法的返回值，但可能线程未执行完成，所以返回 NULL。所以，用轮询的方式，获取方法返回值。缺点是浪费 CPU 周期。

✓ 通过注册回调方法，在构造线程的时候传入回调类，然后，在线程执行过程中调用回调方法。

✓ JDK 1.5 提供了 Future、Callable 和 Executor，Executor 子类 ExecutorService 创建线程，实现 Callable 接口作为回调方法（实现 call() 方法），返回一个 Future 类。

3. 线程同步

当多线程共享资源时，必须考虑同步问题。可以用 synchronized 关键字标注关键对象或方法。但是，同步不仅影响性能，同步得越多越容易造成死锁问题。（死锁：两个线程想要独占某种资源，但是，两者同时占用这种资源的子集的情况）。

同步的替代方法如下。

- 尽可能使用局部变量而不是字段，基本类型传参是值传递，是线程安全的（String 也是安全的，一旦创建不能更改）。
- 构造函数一般不需要考虑线程安全问题。
- 将非线程安全的类作为线程安全类的一个私有字段。
- 注：多线程中使用 System.out 输出，也属于共享资源。

4. 线程的调度

JVM 的线程调度器分为抢占式的（preemptive）和协作式的（cooperative）。由于一个线程长时间占用 CPU，会造成其他线程的饥饿状况。可以通过设置线程的优先级来改变这种情况，但是，在相同优先级的线程中，需要使用如下 6 种类方法手动控制。

- 对 I/O 阻塞。
- 放弃，调用 Thread.yield()，提供给相同优先级的线程 CPU 的使用机会。
- 休眠，sleep()，提供给相同优先级及以下优先级 CPU 的使用机会。
- 连接线程，join()，等待某个指定线程执行结束，或者执行一段时间。
- 等待某个对象，wait()，放弃对一个对象的锁定并暂停（之前的方法并不会放弃资源）。
- 结束，方法返回 return。

6.7 课后习题

一、填空题

1. 调用 Thread 类的________方法可以判断一个线程是否存活。
2. 当某个类实现 Runnable 接口时，需要实现该接口的________方法。
3. Java 语言中线程优先级的默认值是________。

二、选择题

1. Thread 类位于________包中。

A. Java.sql　　B. Java.io

C. Java.lang　　D. Java.util

2. 用户在创建线程时所处的状态是________，在用户使用该线程实例调用 start() 方法之前，线程就处于该状态。

A. 等待状态　　B. 死亡状态

C. 休眠状态　　D. 出生状态

3. 线程同步需要使用________关键字。

A. synchrons　　B. synchrongized

C. implements　　D. extends

4. 下列说法中，错误的是________。

A. 线程调度执行时是按照其优先级的高低顺序执行的

B. 一个线程创建好后即可立即执行

C. 用户程序类可以通过实现 Runnable 接口来定义程序线程

D. 解除处于阻塞状态的线程后，线程便进入就绪状态

三、简答题

1. 什么是进程和线程，两者的区别是什么？
2. 创建线程有哪两种方法，它们各自的优缺点是什么？
3. 简单描述线程的生命周期。
4. 线程的优先级范围是什么？如何实现优先级？
5. 为什么多线程中要引入同步机制？ Java 中如何实现线程的同步？
6. 线程的调度有哪些方法，各有什么功能？

6.8 上机实训

实训一、使用多线程编程，模拟两个以上的银行账户用户同时进行存取操作。要求：

- 余额大于取款金额才可取钱；
- 多人存取完毕后，余额正常。

当一个用户对金额进行修改时，其他用户不可进行修改。

实训二、使用多线程编程，实现素数的判定。待判定的整数由键盘录入后存放在一个列表中，创建 10 个线程从列表中取出整数进行判定，判定的结果存入到另一个列表中，用户可以通过键盘查询判定的结果。

PART07

知识领域7

Java网络编程

知识目标

1. 熟悉网络编程的相关概念，并了解Java网络编程的几种方式。
2. 熟悉InetAddress、Socket、ServerSocket、DatagramSocket、DatagramPacket类的作用和常用方法。

■ 能力目标

1. 熟练通过URL编码获取网站信息和下载文件。
2. 熟练通过TCP编码完成服务器端和客户端的代码编写，实现1对1通信。
3. 熟练使用UDP编码实现在客户端、服务器端之间发送和接收数据报。

■ 素质目标

1. 能够阅读科技文档和撰写分析文档。
2. 能够查阅JDK API。
3. 增强学生团队协作能力。

7.1 应用场景

网络编程就是在两个或两个以上的设备（例如计算机）之间传输数据。网络编程技术是当前一种主流的编程技术，随着互联网的逐步增强以及网络应用程序的大量出现，在实际的开发中网络编程技术获得了大量的使用。

（1）当我们得到一个 URL 对象后，就可以通过它读取指定的 WWW 资源。通过 URL 的方法 openStream()，我们只能从网络上读取数据，通过 URLConnection 可向服务器端的 CGI 程序发送一些数据。WWW、FTP 等标准化的网络服务都是基于 TCP 的，所以本质上讲 URL 编程也是基于 TCP 的一种应用。

（2）客户端上传本地一个文本文件到服务器端，服务器端将客户端上传的文本保存在另一个文本文件中，保存完成后对客户端发送一个“完成保存”的通知，客户端显示服务器返回的信息并断开和服务器的连接，至此整个通信结束。QQ、MSN 都使用了 Socket 相关的技术。

（3）在网络质量不令人十分满意的环境下，UDP 数据包丢失会比较严重。但是由于 UDP 的特性（它不属于连接型协议，因而具有资源消耗小，处理速度快的优点），所以通常音频、视频和普通数据在传送时使用 UDP 较多，因为它们即使偶尔丢失一两个数据包，也不会对接收结果产生太大影响。比如聊天用的 ICQ 和 OICQ 就是使用的 UDP。

Java 所提供的网络功能可大致分为以下三大类。

（1）URL 和 URLConnectio。这是三大类功能中最高级的一种。通过 URL 网络资源表达方式，可以确定了网络上数据的位置。利用 URL 表示和建立，Java 程序可以直接读入网络上的数据，或把自己的数据传送到网络的另一端。

（2）Socket 和 ServerSocket。又称套接字，用于描述 IP 地址和端口。在客户 / 服务器中，当客户机中运行的应用程序（如浏览器）需要访问网络中的服务器时，客户机会被临时分配一个 Socket，然后它会通过服务器的 Socket 向服务器发出请求。换句话讲，网络中的数据发送和接收都是通过 Socket 来完成的，Socket 就像一个文件句柄（读写文件时的一个唯一的顺序号），用户可以通过读写 Socket 来完成客户机 / 服务器直接的通信。

（3）DatagramSocket 和 DatagramPacket。Datagram 是这些功能中最低级的一种。其他网络数据传送方式都假想在程序执行时建立一条安全稳定的通道，但是 Datagram 方式传送数据时，只是把数据的目的地记录在数据包中，然后就直接放在网络上进行传输，系统不保证数据一定能够安全送到，也不能确定什么时候可以送到。也就是说，Datagram 不能保证传送质量。

Java 中有关网络编程方面的功能都定义在 Java.net 包中。程序员所做的事情就是把数据发送到指定的位置，或者接收到指定的数据，这个就属于狭义的网络编程范畴。在发送和接收数据时，大部分的程序设计语言都设计了专门的 API 实现这些功能，程序员只需要调用即可。

7.2 相关知识

7.2.1 网络编程相关知识

网络编程的目的就是指直接或间接地通过网络协议与其他计算机进行通信。网络编程中有两个主要的问题，一个是如何准确地定位网络上一台或多台主机，另一个就是找到主机后如何可靠高效地进行数据传输。在 TCP/IP 中 IP 层主要负责网络主机的定位，数据传输的路由，由 IP 地址可以唯一地确定 Internet 上的一台主机。而 TCP 层则提供面向应用的可靠的或非可靠的数据传输机制，这是网络编程的主要对象，一般不需要关心 IP 层是如何处理数据的。

目前较为流行的网络编程模型是客户机 / 服务器（Client/Server，简称 C/S）结构。即通信双方一方作为服务器等待客户提出请求并予以响应。客户则在需要服务时向服务器提出申请。服务器一般作为守护进程始终运行，监听网络端口，一旦有客户请求，就会启动一个服务进程来响应该客户，同时自己继续监听服务端口，使后来的客户也能及时得到服务。例如 QQ 程序，每个 QQ 用户安装的都是 QQ 客户端程序，而 QQ 服务器端程序则运行在腾讯公司的机房中，为大量的 QQ 用户提供服务。

其实，在运行很多程序时，没有必要使用专用的客户端，而需要使用通用的客户端，例如浏览器，使用浏览器作为客户端的结构被称作浏览器 / 服务器（Browser/Server，B/S）结构。

使用 B/S 结构的程序，在开发时只需要开发服务器端即可，这种结构的优势在于开发的压力比较小，不需要维护客户端。但是这种结构也存在着很多不足，例如浏览器的限制比较大，表现力不强，无法进行系统级操作等。

总之 C/S 结构和 B/S 结构是现在网络编程中常见的两种结构，B/S 结构其实也就是一种特殊的 C/S 结构。

另外简单地介绍一下 P2P（Point to Point）程序，常见的如 BT、电驴等。P2P 程序是一种特殊的程序，一个 P2P 程序中既包含客户端程序，也包含服务器端程序。例如 BT，使用客户端程序部分连接其他的种子（服务器端），而使用服务器端向其他的 BT 客户端传输数据。换一种说法，其实 P2P 程序和手机是一样的，当手机拨打电话时就是使用客户端的作用，而手机处于待机状态时，可以接收到其他用户拨打的电话，则起的就是服务器端的功能，只是一般的手机不能同时使用拨打电话和接听电话的功能，而 P2P 程序实现了该功能。

7.2.2 网络通信方式

在现有的网络中，网络通信的方式主要有以下两种。

① TCP（Transfer Control Protocol，传输控制协议）方式；

② UDP（User Datagram Protocol，用户数据报协议）方式。

TCP 是一种面向连接、可靠的传输协议。通过 TCP 传输，得到的是一个顺序的无差错的数据流。发送方和接收方的成对的两个 Socket 之间必须建立连接，以便在 TCP 的基础上进行通信，当一个 Socket（通常都是 ServerSocket）等待建立连接时，另一个 Socket 可以要求进行连接，一旦这两个 Socket 连接起来，它们就可以进行双向数据传输，双方都可以进行发送或接收操作。

UDP 是一种无连接的、不可靠的协议。每个数据报都是一个独立的信息，包括完整的源地址或目的地址，它在网络上以任何可能的路径传往目的地，因此能否到达目的地，到达目的地的时间以及内容的正确性都是不能被保证的。

在网络通信中，TCP 方式就类似于拨打电话，使用该种方式进行网络通信时，需要建立专门的虚拟连接，然后进行可靠的数据传输，如果数据发送失败，则客户端会自动重发该数据。而 UDP 方式就类似于发送短信，使用这种方式进行网络通信时，不需要建立专门的虚拟连接，传输也不是很可靠，如果发送失败则客户端无法获得。

这两种传输方式都是实际的网络编程中进行使用，重要的数据一般使用 TCP 方式进行传输，而大量的非核心数据则都通过 UDP 方式进行传递。在一些程序中甚至结合使用这两种方式进行数据的传递。

由于 TCP 需要建立专用的虚拟连接以及确认传输是否正确，所以使用 TCP 方式的速度稍微慢一些，而且传输时产生的数据量要比 UDP 稍微大一些。

下面我们对这两种协议做简单比较。具体比较如表 7-1 所示。

表7-1　两种协议比较

特性	TCP	UDP
连接性	面向连接	无连接
可靠性	可靠	不可靠
报文	面向字节流	报文（保留报文的边界）
效率	传输效率低	传输效率高
双工性	全双工	一对一、一对多、多对一、多对多
流量控制	有（滑动窗口）	无（滑动窗口）
拥塞控制	有（慢开始、拥塞避免、快重传、快恢复）	无
应用场景	不定长度的数据，大数据，Telnet，FTP	非核心数据，局域网高可靠性

既然有了保证可靠传输的 TCP，为什么还要非可靠传输的 UDP 呢？主要的原因有两个：一是可靠的传输是要付出代价的，对数据内容正确性的检验必然占用计算机的处理时间和网络的带宽，因此 TCP 传输的效率不如 UDP 高；二是在许多应用中并不需要保证严格的传输可靠性，比如视频会议系统，并不要求音频视频数据绝对正确，只要保证连贯性就可以了，这种情况下显然使用 UDP 会更合理一些。

7.2.3　相关包和类

和网络编程有关的基本 API 位于 Java.net 包中，该包中包含了基本的网络编程实现，是网络编程的基础。该包中既包含基础的网络编程类，也包含封装后的专门处理 Web 相关的处理类，如图 7-1 所示。

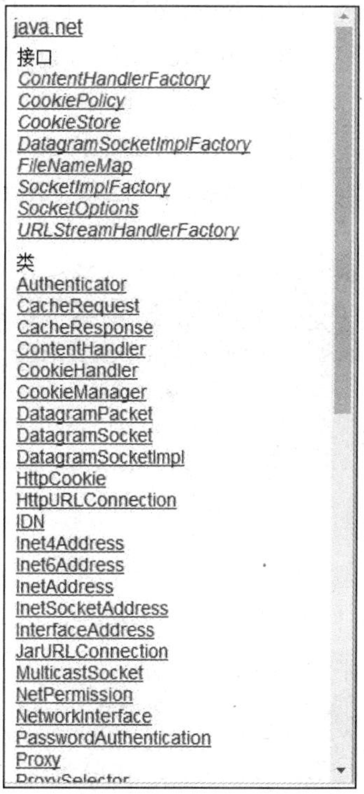

java.net
接口
ContentHandlerFactory
CookiePolicy
CookieStore
DatagramSocketImplFactory
FileNameMap
SocketImplFactory
SocketOptions
URLStreamHandlerFactory
类
Authenticator
CacheRequest
CacheResponse
ContentHandler
CookieHandler
CookieManager
DatagramPacket
DatagramSocket
DatagramSocketImpl
HttpCookie
HttpURLConnection
IDN
Inet4Address
Inet6Address
InetAddress
InetSocketAddress
InterfaceAddress
JarURLConnection
MulticastSocket
NetPermission
NetworkInterface
PasswordAuthentication
Proxy

图7-1　Java.net包

7.3　任务实施

任务一　使用 URL 读取网页内容

1. 任务知识

（1）URL。

统一资源定位器 (Uniform Resource Locator，URL) 是表示 Internet 上某一资源的地址。通过 URL 我们可以访问 Internet 上的各种网络资源，比如最常见的 WWW、FTP 站点。浏览器通过解析给定的 URL 可以在网络上查找相应的文件或其他资源。

（2）URL 的组成。

```
protocol://resourceName
```

协议名（protocol）指明获取资源所使用的传输协议，如 http、ftp、gopher、file 等，资源名（resourceName）则应该是资源的完整地址，包括主机名、端口号、文件名或文件内部的一个引用。例如：

·http://www.sun.com/ 协议名 :// 主机名

http://home.netscape.com/home/welcome.html 协议名 :// 机器名 + 文件名

http://www.gamelan.com:80/Gamelan/network.html#BOTTOM 协议名 :// 机器名 + 端口号 + 文件名 + 内部引用。

（3）URL 类。

Java.net 中实现了类 URL。

① public URL (String spec);

通过一个表示 URL 地址的字符串可以构造一个 URL 对象。

URL urlBase=new URL("http://www. 263.net/")

② public URL(URL context, String spec);

通过基 URL 和相对 URL 构造一个 URL 对象。

URL net263=new URL ("http://www.263.net/");

URL index263=new URL(net263, "index.html")

③ public URL(String protocol, String host, String file);

new URL("http", "www.gamelan.com", "/pages/Gamelan.net. html");

④ public URL(String protocol, String host, int port, String file);

URL gamelan=new URL("http", "www.gamelan.com", 80, "Pages/Gamelan.network.html");

类URL的构造方法都声明抛弃非运行时异常（Malformed URL Exception），因此生成URL对象时，我们必须要对这一例外进行处理，通常是用try-catch语句进行捕获。格式如下：

```
try{
URL myURL= new URL(…)
}catch (MalformedURLException e){
…
}
```

（4）URL 的创建和使用。

```
//创建一个URL的实例
URL baidu =new URL("http://www.baidu.com");
URL url =new URL(baidu,"/index.html?username=tom#test");  //?表示参数，#表示锚点
url.getProtocol();//获取协议
url.getHost();//获取主机
url.getPort();//如果没有指定端口号，根据协议不同使用默认端口。此时getPort()方法的返回值为 -1
url.getPath();//获取文件路径
url.getFile();//文件名，包括文件路径+参数
url.getRef();//相对路径，就是锚点，即#号后面的内容
url.getQuery();//查询字符串，即参数
```

2. 任务需求

得到一个 URL 对象后，通过这个对象读取指定的 WWW 资源。

3. 任务分析

设计类 TestURL，使用 URL 的方法 openStream()，其定义为：InputStream openStream();

方法openSteam()与指定的URL建立连接并返回InputStream类的对象以从这一连接中读取数据。

4．任务实现

TestURL.java

```
package com.daiinfo.seniorjava.ken7.implement;
import java.io.BufferedReader;
import java.io.InputStream;
import java.io.InputStreamReader;
import java.net.URL;
public class TestURL {
    public static void main(String[] args) {
        try {
            String s_url = "https://www.baidu.com/";
            URL url = new URL(s_url);
            System.out.println(url.getAuthority());// 获得此URL的授权部分
            System.out.println(url.getContent());// 获得此URL的内容
            System.out.println(url.getDefaultPort());// 获得与此URL关联协议的默认端口号
            System.out.println(url.getFile());// 获得此URL的文件名
            System.out.println(url.getHost());// 获得此URL的主机名（如果适用）
            System.out.println(url.getPath());// 获得此URL的路径部分
            System.out.println(url.getPort());// 获得此URL的端口号
            System.out.println(url.getProtocol());// 获得此URL的协议名称
            InputStream in = url.openStream();// 返回读入网站内容的字节输入流
            // 上一句等同于下面两句
            // URLConnection conn = url.openConnection();
            // InputStream in = conn.getInputStream();
            // 转换成带缓冲的字符输入流
            BufferedReader reader = new BufferedReader(
                    new InputStreamReader(in));
            String msg = null;
            // 读取内容，直到为空
            while ((msg = reader.readLine()) != null) {
                // 把内容输出到控制台
                System.out.println(msg);
            }
        } catch (Exception e) {
            e.printStackTrace();
        }
    }
}
```

运行结果如图 7-2 所示（部分内容省略）。

```
www.baidu.com
sun.net.www.protocol.http.HttpURLConnection$HttpInputStream@df8388a
443
/
www.baidu.com
/
-1
https
<!DOCTYPE html>
<!--STATUS OK--><html> <head><meta http-equiv=content-type content=text/html;charset=utf-8>
                </script> <a href=//www.baidu.com/more/ name=tj_briicon class=bri style="di
```

图7-2　运行结果

任务二　基于 TCP 编程

1. 任务知识

（1）Socket 通信。

网络上的两个程序通过一个双向的通信连接实现数据的交换，这个双向链路的一端称为一个Socket。Socket 通常用来实现客户机和服务器的连接。Socket 是 TCP/IP 的一个十分流行的编程界面，一个 Socket 由一个 IP 地址和一个端口号唯一确定。

在传统的 UNIX 环境下可以操作 TCP/IP 的接口不止 Socket 一个，Socket 所支持的协议种类也不光 TCP/IP 一种，因此两者之间是没有必然联系的。在 Java 环境下，Socket 编程主要是指基于 TCP/IP 的网络编程。

（2）相关类。

① Socket 类。

Socket（套接字）：代表一个 IP 和 port 组合，也就是两个程序的连接。

Socket 的作用：获取输入输出流。

类的作用：用于向服务端发送请求，通过 Ip 和 port 请求建立连接。

构造方法：通过 IP 和 port 请求与 B 程序建立连接。

通过 IP 地址（字符串）和端口构造：

```
Socket ss = new Socket("127.0.0.1",7890);
```

通过 InetAddress 对象和端口构造：

```
InetAddress addr = InetAddress.getByName("128.0.43.197");
Socket ss = new Socket(addr,7890);
```

常用方法：

```
InputStream is = ss.getInputStream();
OutputStream os = ss.getOutputStream();
ss.close();
```

② ServerSocket 类。

ServerSocket 用来描述网络服务端。

类的作用：创建一个网络服务，等待客户端连接。

构造方法：创建一个服务，并占用一个端口号。

```
ServerSocket server = new ServerSocket(7890);
```

常用方法：通过 ServerSocket 对象，调用 accept 方法，让服务处于等待状态，并获得 Socket。

```
Socket s = server.accept();
```

网络通信结束以后关闭服务：

```
server.close();
```

③ InetAddress 类。

该类的功能是代表一个 IP 地址，并且将 IP 地址和域名相关的操作方法包含在该类的内部。

```
//使用域名创建对象
InetAddress inet1 = InetAddress.getByName("www.163.com");
//使用IP创建对象
InetAddress inet2 = InetAddress.getByName("127.0.0.1");
```

```
//获得本机地址对象
InetAddress inet3 = InetAddress.getLocalHost();
//获得对象中存储的域名
String host = inet3.getHostName();
//获得对象中存储的IP
String ip = inet3.getHostAddress();
```

（3）Socket 通信的一般过程。

使用 Socket 进行 Client/Server 程序设计的一般连接过程是：Server 端 Listen（监听）某个端口是否有连接请求，Client 端向 Server 端发出 Connect（连接）请求，Server 端向 Client 端发回 Accept（接受）消息，一个连接就建立起来了。Server 端和 Client 端都可以通过 Send、Write 等方法与对方通信。

对于一个功能齐全的 Socket，都要包含以下基本结构，如图 7-3 所示。其工作过程包含以下 4 个基本的步骤。

- 创建 Socket；
- 打开连接到 Socket 的输入 / 出流；
- 按照一定的协议对 Socket 进行读 / 写操作；
- 关闭 Socket。

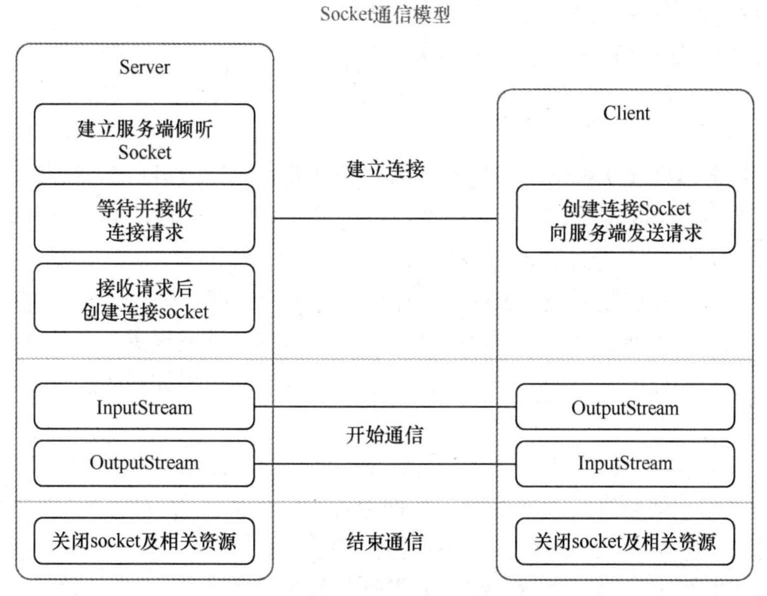

图7-3　Socket通信模型

（4）客户机 / 服务器编程步骤。

① 客户端网络编程步骤。

客户端（Client）是指网络编程中首先发起连接的程序，客户端一般实现程序界面和基本逻辑实现，在进行实际的客户端编程时，无论客户端复杂还是简单，以及客户端实现的方式，客户端的编程都主要由下面几个步骤实现。

- 创建 Socket 对象，指明需要连接的服务器的地址和端口号；
- 连接建立后，通过输出流向服务器端发送请求信息；
- 通过输入流获取服务器响应的信息；

- 关闭相应资源。

建立网络连接：客户端网络编程的第一步都是建立网络连接。在建立网络连接时需要指定连接到的服务器的 IP 地址和端口号，建立完成以后，会形成一条虚拟的连接，后续的操作就可以通过该连接实现数据交换了。

交换数据：连接建立以后，就可以通过这个连接交换数据了。交换数据严格按照请求响应模型进行，由客户端发送一个请求数据到服务器，服务器反馈一个响应数据给客户端，如果客户端不发送请求则服务器端就不响应。根据逻辑需要，可以多次交换数据，但是还是必须遵循请求响应模型。

关闭网络连接：在数据交换完成以后，关闭网络连接，释放程序占用的端口、内存等系统资源，结束网络编程。

最基本的步骤一般都是这几个步骤，在实际实现时，步骤 2 会出现重复，在进行代码组织时，由于网络编程是比较耗时的操作，所以一般开启专门的现场进行网络通信。

② 服务器端网络编程步骤。

服务器端（Server）是指在网络编程中被动等待连接的程序，服务器端一般实现程序的核心逻辑以及数据存储等核心功能。服务器端的编程步骤和客户端不同，是由下面几个步骤实现的。

- 创建 ServerSocket 对象，绑定监听端口；
- 通过 accept() 方法监听客户端请求；
- 连接建立后，通过输入流读取客户端发送的请求信息；
- 通过输出流向客户端发送响应信息；
- 关闭相关资源。

监听端口：服务器端属于被动等待连接，所以服务器端启动以后，不需要发起连接，而只需要监听本地计算机的某个固定端口即可。这个端口就是服务器端开放给客户端的端口，服务器端程序运行的本地计算机的 IP 地址就是服务器端程序的 IP 地址。

获得连接：当客户端连接到服务器端时，服务器端就可以获得一个连接，这个连接包含客户端的信息，例如客户端 IP 地址等等，服务器端和客户端也通过该连接进行数据交换。

一般在服务器端编程中，当获得连接时，需要开启专门的线程处理该连接，每个连接都由独立的线程实现。

交换数据：服务器端通过获得的连接进行数据交换。服务器端的数据交换步骤是首先接收客户端发送过来的数据，然后进行逻辑处理，再把处理以后的结果数据发送给客户端。简单来说，就是先接收再发送，这个和客户端的数据交换顺序不同。

关闭连接：当服务器程序关闭时，需要关闭服务器端，通过关闭服务器端使得服务器监听的端口以及占用的内存可以释放出来，实现了连接的关闭。

其实服务器端编程的模型和呼叫中心的实现是类似的，例如移动的客服电话 10086 就是典型的呼叫中心，当一个用户拨打 10086 时，转接给一个专门的客服人员，由该客服实现该用户的问题解决，当另外一个用户拨打 10086 时，则转接给另一个客服，实现问题解决，依次类推。在服务器端编程时，10086 这个电话号码就类似于服务器端的端口号码，每个用户就相当于一个客户端程序，每个客服人员就相当于服务器端启动的专门和客户端连接的线程，每个线程都是独立进行交互的。

这就是服务器端编程的模型，只是 TCP 方式是需要建立连接的，对于服务器端的压力比较大，而 UDP 是不需要建立连接的，对于服务器端的压力比较小罢了。

2. 任务需求

完成简单的一对一通信。

3. 任务分析

本任务的类图如图 7-4 所示。

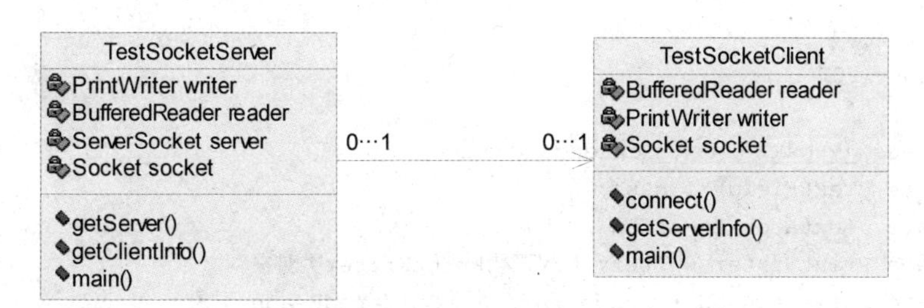

图7-4　一对一通信类图

时序图如图 7-5 所示。

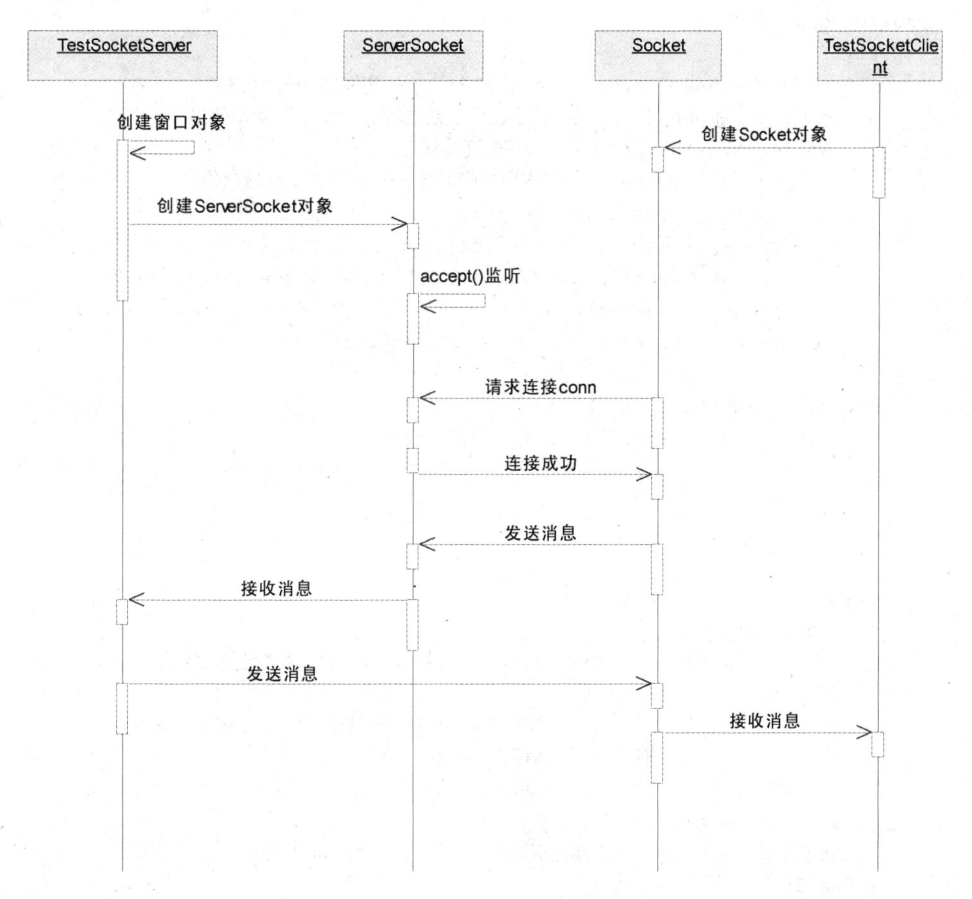

图7-5　一对一通信时序图

4. 任务实现

TestSocketServer.java

```
package com.daiinfo.seniorjava.ken7.implement;
import java.awt.*;
```

```
import java.awt.event.*;
import java.io.*;
import java.net.*;
import javax.swing.*;
/**
 * 服务器端
 * @author 戴远泉
 * @version 1.0
 */
public class TestSocketServer extends JFrame {
    private JTextField tf_send;
    private JTextArea ta_info;
    private PrintWriter writer; // 声明PrintWriter类对象
    private BufferedReader reader; // 声明BufferedReader对象
    private ServerSocket server; // 声明ServerSocket对象
    private Socket socket; // 声明Socket对象socket

    public void getServer() {
        try {
            server = new ServerSocket(1978); // 实例化Socket对象
            ta_info.append("服务器套接字已经创建成功\n"); // 输出信息
            while (true) { // 如果套接字是连接状态
                ta_info.append("等待客户机的连接……\n"); // 输出信息
                socket = server.accept(); // 实例化Socket对象
                reader = new BufferedReader(new InputStreamReader(socket
                        .getInputStream())); // 实例化BufferedReader对象
                writer = new PrintWriter(socket.getOutputStream(), true);
                getClientInfo(); // 调用getClientInfo()方法
            }
        } catch (Exception e) {
            e.printStackTrace(); // 输出异常信息
        }
    }

    private void getClientInfo() {
        try {
            while (true) {
                String line = reader.readLine();// 读取客户端发送的信息
                if (line != null)
                    ta_info.append("接收到客户机发送的信息: " + line + "\n");
                    // 显示客户端发送的信息
            }
        } catch (Exception e) {
            ta_info.append("客户端已退出。\n"); // 输出异常信息
        } finally {
            try {
                if (reader != null) {
                    reader.close();// 关闭流
                }
                if (socket != null) {
                    socket.close(); // 关闭套接字
```

```
            }
        } catch (IOException e) {
            e.printStackTrace();
        }
    }
}

public static void main(String[] args) { // 主方法
    TestSocketServer frame = new TestSocketServer(); // 创建本类对象
    frame.setVisible(true);
    frame.getServer(); // 调用方法
}

public TestSocketServer() {
    super();
    setTitle("服务器端程序");
    setDefaultCloseOperation(JFrame.EXIT_ON_CLOSE);
    setBounds(100, 100, 379, 260);

    final JScrollPane scrollPane = new JScrollPane();
    getContentPane().add(scrollPane, BorderLayout.CENTER);

    ta_info = new JTextArea();
    scrollPane.setViewportView(ta_info);

    final JPanel panel = new JPanel();
    getContentPane().add(panel, BorderLayout.SOUTH);

    final JLabel label = new JLabel();
    label.setText("服务器发送的信息: ");
    panel.add(label);

    tf_send = new JTextField();
    tf_send.setPreferredSize(new Dimension(150, 25));
    panel.add(tf_send);
    final JButton button = new JButton();
    button.addActionListener(new ActionListener() {
        public void actionPerformed(final ActionEvent e) {
            writer.println(tf_send.getText()); // 将文本框中信息写入流
            ta_info.append("服务器发送的信息是: " + tf_send.getText() + "\n");
            // 将文本框中信息显示在文本域中
            tf_send.setText(""); // 将文本框清空
        }
    });
    button.setText("发送");
    panel.add(button);
    final JPanel panel_1 = new JPanel();
    getContentPane().add(panel_1, BorderLayout.NORTH);
    final JLabel label_1 = new JLabel();
    label_1.setForeground(new Color(0, 0, 255));
    label_1.setFont(new Font("", Font.BOLD, 22));
```

```
            label_1.setText("一对一通信——服务器端程序");
            panel_1.add(label_1);
        }
    }
```

TestSocketClient.java

```
package com.daiinfo.seniorjava.ken7.implement;
import java.awt.*;
import java.awt.event.*;
import java.io.*;
import java.net.*;
import javax.swing.*;
/**
 * 客户端
 * @author 戴远泉
 * @version 1.0s
 */
public class TestSocketClient extends JFrame {
    private PrintWriter writer; // 声明PrintWriter类对象
    private BufferedReader reader; // 声明BufferedReader对象
    private Socket socket; // 声明Socket对象
    private JTextArea ta_info; // 创建JtextArea对象
    private JTextField tf_send; // 创建JtextField对象
    private void connect() { // 连接套接字方法
        ta_info.append("尝试连接……\n"); // 文本域中的信息
        try { // 捕捉异常
            socket = new Socket("127.0.0.1", 1978); // 实例化Socket对象
            while (true) {
                writer = new PrintWriter(socket.getOutputStream(), true);
                reader = new BufferedReader(new InputStreamReader(socket
                        .getInputStream())); // 实例化BufferedReader对象
                ta_info.append("完成连接。\n"); // 文本域中提示信息
                getServerInfo();
            }
        } catch (Exception e) {
            e.printStackTrace(); // 输出异常信息
        }
    }

    public static void main(String[] args) { // 主方法
        TestSocketClient clien = new TestSocketClient(); // 创建本例对象
        clien.setVisible(true); // 将窗体显示
        clien.connect(); // 调用连接方法
    }

    private void getServerInfo() {
        try {
            while (true) {
                if (reader != null) {
                    String line = reader.readLine();// 读取服务器发送的信息
                    if (line != null)
```

```
                    ta_info.append("接收到服务器发送的信息: " + line + "\n");
                    // 显示服务器端发送的信息
                }
            }
        } catch (Exception e) {
            e.printStackTrace();
        } finally {
            try {
                if (reader != null) {
                    reader.close();// 关闭流
                }
                if (socket != null) {
                    socket.close(); // 关闭套接字
                }
            } catch (IOException e) {
                e.printStackTrace();
            }
        }
    }

    /**
     * Create the frame
     */
    public TestSocketClient() {
        super();
        setTitle("客户端程序");
        setBounds(100, 100, 361, 257);
        setDefaultCloseOperation(JFrame.EXIT_ON_CLOSE);
        final JPanel panel = new JPanel();
        getContentPane().add(panel, BorderLayout.NORTH);
        final JLabel label = new JLabel();
        label.setForeground(new Color(0, 0, 255));
        label.setFont(new Font("", Font.BOLD, 22));
        label.setText("一对一通信——客户端程序");
        panel.add(label);
        final JPanel panel_1 = new JPanel();
        getContentPane().add(panel_1, BorderLayout.SOUTH);
        final JLabel label_1 = new JLabel();
        label_1.setText("客户端发送的信息: ");
        panel_1.add(label_1);
        tf_send = new JTextField();
        tf_send.setPreferredSize(new Dimension(140, 25));
        panel_1.add(tf_send);
        final JButton button = new JButton();
        button.addActionListener(new ActionListener() {
            public void actionPerformed(final ActionEvent e) {
                writer.println(tf_send.getText()); // 将文本框中信息写入流
                ta_info.append("客户端发送的信息是: " + tf_send.getText()
                        + "\n"); // 将文本框中信息显示在文本域中
                tf_send.setText(""); // 将文本框清空
            }
```

```
            });
            button.setText("发送");
            panel_1.add(button);
            final JScrollPane scrollPane = new JScrollPane();
            getContentPane().add(scrollPane, BorderLayout.CENTER);
            ta_info = new JTextArea();
            scrollPane.setViewportView(ta_info);
        }
    }
```

运行结果如图 7-6 和图 7-7 所示。

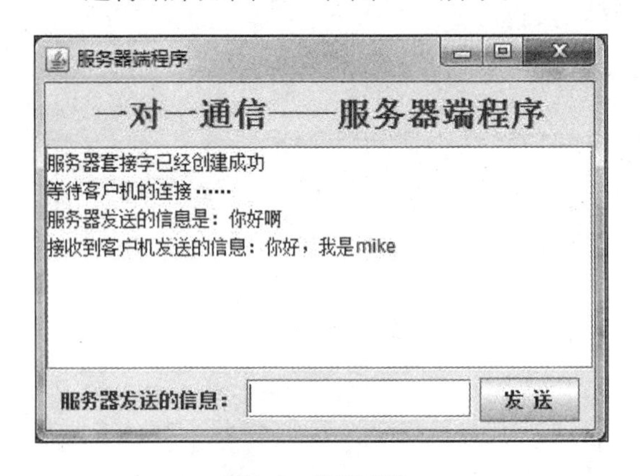

图7-6　服务器端

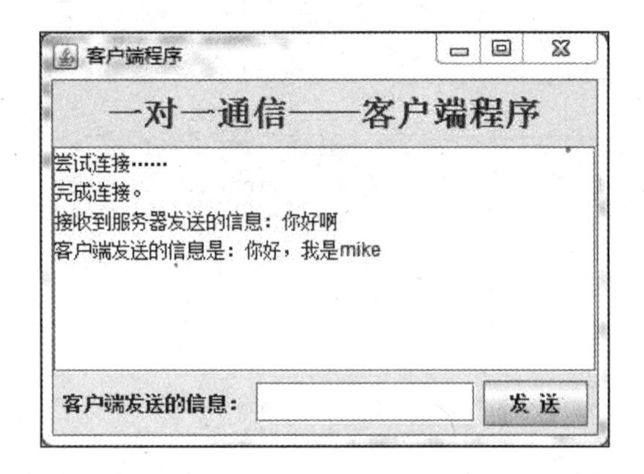

图7-7　客户端

在选择端口时，必须小心。每一个端口提供一种特定的服务，只有给出正确的端口，才能获得相应的服务。0～1023的端口号为系统所保留，例如http服务的端口号为80,telnet服务的端口号为21,ftp服务的端口号为23，所以我们在选择端口号时，最好选择一个大于1023的数以防止发生冲突。

在创建 Socket 时如果发生错误，将产生 IOException，在程序中必须对之做出处理。所以在创建 Socket 或 ServerSocket 时必须捕获或抛出例外。

任务三　基于 UDP 编程

1. 任务知识

（1）UDP。

UDP 方式的网络编程也在 Java 语言中获得了良好的支持，由于其在传输数据的过程中不需要建立专用的连接等特点，所以在 Java API 中设计的实现结构和 TCP 方式不太一样。

UDP 是无连接的、不可靠的、无序的，速度快。进行数据传输时，首先将要传输的数据定义成数据报（Datagram），大小限制在 64k，在数据报中指明数据所要达到的 Socket（主机地址和端口号），然后再将数据报发送出去。

（2）相关类。

① DatagramPacket 类。此类表示数据报包。

类的作用：一个 DatagramPacket 对象表示一个数据包，可以用来封装 UDP 传输的数据。

构造方法：

```
DatagramPacket(byte[] bs, int length)
DatagramPacket(byte[] bs, int length,InetAddress addr, int port)
```

创建对象：

```
byte bs[] = new byte[1024];
InetAddress address = InetAddress.getByName("192.168.1.12");
//如果是发送的数据包，则需要指定地址和端口号，如果是接收的，则不需要
DatagramPacket dp_send = new DatagramPacket(bs, bs.length,address,8989);
DatagramPacket dp_receive = new DatagramPacket(bs, bs.length);
```

常用方法：

```
//得到接收的数据
dp_receive.getData();
//得到数据长度
dp_receive.getLength();
```

② DatagramSocket 类。

用于发送和接收 UDP 数据，和 Socket 没有关系。

类的作用：一个 DatagramSocket 对象，既可以用来发送数据报，也可以接收数据报。

构造方法：DatagramSocket()——发送数据报创建的对象，通常使用无参构造器。

DatagramSocket (int port)——接收数据通常使用有参构造器。

创建对象：

```
//DatagramSocket用来发送数据包
DatagramSocket ds_send = new DatagramSocket();
//DatagramSocket用来接收数据包
DatagramSocket ds_receive = new DatagramSocket(8989);
```

常用方法：

```
//发送数据包
ds_send.send(dp_send);
//接收数据包
ds_receive.receive(dp_receive);
```

（3）UDP 编程步骤。

服务器端实现步骤如下。

① 创建 DatagramSocket，指定端口号；

② 创建 DatagramPacket ；

③ 接受客户端发送的数据信息；

④ 读取数据。

客户端实现步骤如下。

① 定义发送信息；

② 创建 DatagramPacket，包含将要发送的信息；

③ 创建 DatagramSocket ；

④ 发送数据。

（4）注意问题。

① 多线程的优先级问题。

根据实际的经验，适当地降低优先级，否则可能会出现程序运行效率低的情况。

② 是否关闭输出流和输入流。

对于同一个 Socket，如果关闭了输出流，则与该输出流关联的 Socket 也会被关闭，所以一般不用关闭流，直接关闭 Socket 即可。

③ 使用 TCP 通信传输对象，IO 中序列化部分。

④ Socket 编程传递文件，IO 流部分。

2. 任务需求

发送端向接收端发送信息。

3. 任务实现

TestUDPReceiver.java

```
package com.daiinfo.seniorjava.ken7.implement;
import java.io.IOException;
import java.net.DatagramPacket;
import java.net.DatagramSocket;
import java.net.InetAddress;
import java.net.SocketException;
public class TestUDPReceiver {
    public static void main(String[] args) {
        // ①创建服务器端DatagramSocket，指定端口
        DatagramSocket socket = null;
        try {
            socket = new DatagramSocket(10010);
        } catch (SocketException e1) {
            // TODO Auto-generated catch block
            e1.printStackTrace();
        }
        // ②创建数据报，用于接受客户端发送的数据
        byte[] data = new byte[1024];//
        DatagramPacket packet = new DatagramPacket(data, data.length);
        // ③接收客户端发送的数据
        try {
            socket.receive(packet);// 此方法在接收数据报之前会阻塞
        } catch (IOException e) {
            // TODO Auto-generated catch block
            e.printStackTrace();
        }
        // ④读取数据
        String info = new String(data, 0, data.length);
        System.out.println("我是服务器，客户端告诉我" + info);
        // =========================================================
        // 向客户端响应数据
        // ①定义客户端的地址、端口号、数据
        InetAddress address = packet.getAddress();
        int port = packet.getPort();
        byte[] data2 = "欢迎您！".getBytes();
        // ②创建数据报，包含响应的数据信息
        DatagramPacket packet2 = new DatagramPacket(data2, data2.length,
                address, port);
```

```
        // ③响应客户端
        try {
            socket.send(packet2);
        } catch (IOException e) {
            // TODO Auto-generated catch block
            e.printStackTrace();
        }
        // ④关闭资源
        socket.close();
    }
}
```

TestUDPSender.java

```
package com.daiinfo.seniorjava.ken7.implement;
import java.io.IOException;
import java.net.DatagramPacket;
import java.net.DatagramSocket;
import java.net.InetAddress;
import java.net.SocketException;
import java.net.UnknownHostException;
public class TestUDPSender {
    public static void main(String[] args) {
        // ①定义服务器的地址、端口号、数据
        InetAddress address = null;
        try {
            address = InetAddress.getByName("localhost");
        } catch (UnknownHostException e) {
            // TODO Auto-generated catch block
            e.printStackTrace();
        }
        int port = 10010;
        byte[] data = "用户名: admin;密码: 123".getBytes();
        // ②创建数据报，包含发送的数据信息
        DatagramPacket packet = new DatagramPacket(data, data.length, address,
                port);
        // ③创建DatagramSocket对象
        DatagramSocket socket = null;
        try {
            socket = new DatagramSocket();
        } catch (SocketException e) {
            // TODO Auto-generated catch block
            e.printStackTrace();
        }
        // ④向服务器发送数据
        try {
            socket.send(packet);
        } catch (IOException e) {
            // TODO Auto-generated catch block
            e.printStackTrace();
        }
        // 接收服务器端响应数据
```

```
        // ======================================
        // ①创建数据报，用于接收服务器端响应数据
        byte[] data2 = new byte[1024];
        DatagramPacket packet2 = new DatagramPacket(data2, data2.length);
        // ②接收服务器响应的数据
        try {
            socket.receive(packet2);
        } catch (IOException e) {
            // TODO Auto-generated catch block
            e.printStackTrace();
        }
        String raply = new String(data2, 0, packet2.getLength());
        System.out.println("我是客户端，服务器说：" + raply);
        // ③关闭资源
        socket.close();
    }
}
```

运行结果如图 7-8 所示。

```
Markers  Properties  Servers  Data Source Explorer  Snippets  Console  Search
<terminated> TestUDPReceiver [Java Application] C:\Java\jdk1.7.0_67\bin\javaw.exe (2017年8月17日 下午3:10:08)
我是服务器，客户端告诉我用户名：admin;密码：123
```

图7-8 运行结果

7.4 拓展知识

为了一步一步地掌握网络编程，下面再研究网络编程中的两个基本问题，我们通过解决这两个问题将会对网络编程的认识深入一层。

1. 如何复用 Socket 连接

在前面的示例中，客户端中建立了一次连接，只发送一次数据就关闭了，这就相当于拨打电话时，电话打通了只对话一次就关闭了，其实更加常用的应该是拨通一次电话以后多次对话，这就是复用客户端连接。

那么如何实现建立一次连接，进行多次数据交换呢？其实很简单，建立连接以后，将数据交换的逻辑写到一个循环中就可以了。这样只要循环不结束则连接就不会被关闭。按照这种思路，可以改造一下上面的代码，让该程序可以在建立连接一次以后，发送 3 次数据，当然这里的次数也可以是多次，示例代码如下：

```
package com.daiinfo.seniorjava.ken7.prolongation;
import java.io.*;
import java.net.*;
/**
 * 复用连接的Socket客户端功能为：发送字符串“Hello”到服务器端，并打印出服务器端的反馈
 */
public class TestMulSocketClient {
    public static void main(String[] args) {
```

```
        Socket socket = null;
        InputStream is = null;
        OutputStream os = null;
        // 服务器端IP地址
        String serverIP = "127.0.0.1";
        // 服务器端端口号
        int port = 10000;
        // 发送内容
        String data[] = { "First", "Second", "Third" };
        try {
            // 建立连接
            socket = new Socket(serverIP, port);
            // 初始化流
            os = socket.getOutputStream();
            is = socket.getInputStream();
            byte[] b = new byte[1024];
            for (int i = 0; i < data.length; i++) {
                // 发送数据
                os.write(data[i].getBytes());
                // 接收数据
                int n = is.read(b);
                // 输出反馈数据
                System.out.println("服务器反馈: " + new String(b, 0, n));
            }
        } catch (Exception e) {
            e.printStackTrace(); // 打印异常信息
        } finally {
            try {
                // 关闭流和连接
                is.close();
                os.close();
                socket.close();
            } catch (Exception e2) {
            }
        }
    }
}
```

该示例程序和前面的代码相比，将数据交换部分的逻辑写在一个 for 循环的内容，这样就可以建立一次连接，依次将 data 数组中的数据按照顺序发送给服务器端了。

如果还是使用前面示例代码中的服务器端程序运行该程序，则该程序的结果是：

```
java.net.SocketException: Software caused connection abort: recv failed
at java.net.SocketInputStream.socketRead0(Native Method)
at java.net.SocketInputStream.read(SocketInputStream.java:129)
at java.net.SocketInputStream.read(SocketInputStream.java:90)
at tcp.MulSocketClient.main(MulSocketClient.java:30)
```

服务器反馈：First。

显然，客户端在实际运行时出现了异常，出现异常的原因是什么呢？如果仔细阅读前面的代码，应该还记得前面示例代码中的服务器端是对话一次数据以后就关闭了连接，如果服务器端程序关闭了，客户端继续发送数据肯定会出现异常，这就是出现该问题的原因。

按照客户端实现的逻辑，也可以复用服务器端的连接，实现的原理也是将服务器端的数据交换逻辑写在循环中即可，按照该种思路改造以后的服务器端代码为：

```
package com.daiinfo.seniorjava.ken7.prolongation;
import java.io.*;
import java.net.*;
/**
 * 复用连接的echo服务器功能：将客户端发送的内容反馈给客户端
 */
public class TestMulSocketServer {
    public static void main(String[] args) {
        ServerSocket serverSocket = null;
        Socket socket = null;
        OutputStream os = null;
        InputStream is = null;
        // 监听端口号
        int port = 10000;
        try {
            // 建立连接
            serverSocket = new ServerSocket(port);
            System.out.println("服务器已启动：");
            // 获得连接
            socket = serverSocket.accept();
            // 初始化流
            is = socket.getInputStream();
            os = socket.getOutputStream();
            byte[] b = new byte[1024];
            for (int i = 0; i < 3; i++) {
                int n = is.read(b);
                // 输出
                System.out.println("客户端发送内容为：" + new String(b, 0, n));
                // 向客户端发送反馈内容
                os.write(b, 0, n);
            }
        } catch (Exception e) {
            e.printStackTrace();
        } finally {
            try {
                // 关闭流和连接
                os.close();
                is.close();
                socket.close();
                serverSocket.close();
            } catch (Exception e) {
            }
        }
    }
}
```

运行结果如图 7-9 所示。

Markers Properties Servers Data Source Explorer Snippets Console Search
<terminated> TestMulSocketClient [Java Application] C:\Java\jdk1.7.0_67\bin\javaw.exe (2017年8月17日 下午3:35:12)
服务器反馈：First
服务器反馈：Second
服务器反馈：Third

图7-9 运行结果

2. 如何使服务器端支持多个客户端同时工作

前面介绍的服务器端程序，只是实现了概念上的服务器端，离实际的服务器端程序结构距离还很遥远，如果需要让服务器端能够实际使用，那么最需要解决的问题就是——如何支持多个客户端同时工作。

一个服务器端一般都需要同时为多个客户端提供通信，如果需要同时支持多个客户端，则必须使用前面介绍的线程的概念。简单来说，也就是当服务器端接收到一个连接时，启动一个专门的线程处理和该客户端的通信。

按照这个思路改写的服务端示例程序将由两个部分组成，TestMulThreadSocketServer 类实现服务器端控制，实现接收客户端连接，然后开启专门的逻辑线程处理该连接。LogicThread 类实现对于一个客户端连接的逻辑处理，将处理的逻辑放置在该类的 run 方法中。该示例的代码实现为：

```
package com.daiinfo.seniorjava.ken7.prolongation;
import java.net.ServerSocket;
import java.net.Socket;
/**
 * 支持多客户端的服务器端实现
 */
public class TestMulThreadSocketServer {
    public static void main(String[] args) {
        ServerSocket serverSocket = null;
        Socket socket = null;
        // 监听端口号
        int port = 10000;
        try {
            // 建立连接
            serverSocket = new ServerSocket(port);
            System.out.println("服务器已启动: ");
            while (true) {
                // 获得连接
                socket = serverSocket.accept();
                // 启动线程
                new LogicThread(socket);
            }
        } catch (Exception e) {
            e.printStackTrace();
        } finally {
            try {
                // 关闭连接
                serverSocket.close();
            } catch (Exception e) {
            }
        }
```

```
    }
}
```

在该示例代码中，实现了一个 while 形式的死循环，由于 accept 方法是阻塞方法，所以当客户端连接未到达时，将阻塞该程序的执行，当客户端到达时接收该连接，并启动一个新的 LogicThread 线程处理该连接，然后按照循环的执行流程，继续等待下一个客户端连接。这样当任何一个客户端连接到达时，都开启一个专门的线程处理，通过多个线程支持多个客户端同时处理。

```
package com.daiinfo.seniorjava.ken7.prolongation;
import java.io.*;
import java.net.*;
/**
 * 服务器端逻辑线程
 */
public class LogicThread extends Thread {
        Socket socket;
        InputStream is;
        OutputStream os;
        public LogicThread(Socket socket){
                this.socket = socket;
                start(); //启动线程
        }

        public void run(){
                byte[] b = new byte[1024];
                try{
                        //初始化流
                        os = socket.getOutputStream();
                        is = socket.getInputStream();
                        for(int i = 0;i < 3;i++){
                                //读取数据
                                int n = is.read(b);
                                //逻辑处理
                                byte[] response = logic(b,0,n);
                                //反馈数据
                                os.write(response);
                        }
                }catch(Exception e){
                        e.printStackTrace();
                }finally{
                        close();
                }
        }

        /**
         * 关闭流和连接
         */
        private void close(){
                try{
                        //关闭流和连接
                        os.close();
                        is.close();
```

```
                    socket.close();
                }catch(Exception e){}
        }

        /**
         * 逻辑处理方法，实现echo逻辑
         * @param b 客户端发送数据缓冲区
         * @param off 起始下标
         * @param len 有效数据长度
         * @return
         */
        private byte[] logic(byte[] b,int off,int len){
                byte[] response = new byte[len];
                //将有效数据复制到数组response中
                System.arraycopy(b, 0, response, 0, len);
                return response;
        }
}s
```

在该示例代码中，每次使用一个连接对象构造该线程，该连接对象就是该线程需要处理的连接，在线程构造完成以后，该线程就被启动起来了，然后在 run 方法内部对客户端连接进行处理，数据交换的逻辑和前面的示例代码一致，只是这里将接收到客户端发送过来的数据并进行处理的逻辑封装成了 logic 方法，按照前面介绍的 IO 编程的内容，客户端发送过来的内容存储在数组 b 的起始下标为 0，长度为 *n* 个中，这些数据是客户端发送过来的有效数据，将有效的数据传递给 logic 方法，logic 方法实现的是 echo 服务的逻辑，也就是将客户端发送的有效数据形成新的 response 数组，并作为返回值反馈。

在线程中将 logic 方法的返回值反馈给客户端，这样就完成了服务器端的逻辑处理模拟，其他的实现和前面的介绍类似，这里就不再重复了。

这里的示例还只是基础的服务器端实现，在实际的服务器端实现中，由于硬件和端口数的限制，所以不能无限制地创建线程对象，而且频繁地创建线程对象效率也比较低，所以程序中都实现了线程池来提高程序的执行效率。

这里简单介绍一下线程池的概念，线程池（Thread pool）是池技术的一种，就是在程序启动时首先把需要个数的线程对象创建好，例如创建 5000 个线程对象，然后当客户端连接到达时从池中取出一个已经创建完成的线程对象使用即可。当客户端连接关闭以后，将该线程对象重新放入到线程池中供其他的客户端重复使用，这样可以提高程序的执行速度，优化程序对于内存的占用等。

7.5 拓展训练

1. 任务需求

完成 UDP 多播传输。

2. 任务实现

MultiSender.java

```
package com.daiinfo.seniorjava.ken7.prolongation;
import java.net.*;
import java.util.Scanner;
```

```
public class MultiSender {
    public static void main(String[] args) {
        try {
            // ①载入信息，从键盘输入
            Scanner sc = new Scanner(System.in);
            System.out.println("请输入要发送的信息：");
            String msg = sc.nextLine();
            // ②创建数据报
            InetAddress address = InetAddress.getByName("228.5.6.7");
            byte[] bs = msg.getBytes();
            DatagramPacket dp = new DatagramPacket(bs, bs.length, address, 8989);
            // ③通过MultiSocket发送数据报
            MulticastSocket ms = new MulticastSocket();
            ms.joinGroup(address);
            ms.send(dp);
        } catch (Exception e) {
            e.printStackTrace();
        }
    }
}
```

7.6 课后小结

1. URL 传输

统一资源定位器（Uniform Resoure Locator，URL）是 WWW 页的地址，它从左到右由下述部分组成。

- Internet 资源类型（scheme）：指出 WWW 客户程序用来操作的工具。如“http://”表示 WWW 服务器，“ftp://”表示 FTP 服务器，“gopher://”表示 Gopher 服务器，而“new:”表示 Newgroup 新闻组。
- 服务器地址（host）：指出 WWW 页所在的服务器域名。
- 端口（port）：有时（并非总是这样），对某些资源的访问来说，需给相应的服务器提供端口号。
- 路径（path）：指明服务器上某资源的位置（其格式与 DOS 系统中的格式一样，通常由目录 / 子目录 / 文件名这样的结构组成）。与端口一样，路径并非总是需要的。

URL 地址格式排列为：scheme://host:port/path，例如 http://www.sohu。

2. TCP 传输

两个端点的建立连接后会有一个传输数据的通道，这通道称为流，而且是建立在网络基础上的流，称之为 Socket 流。该流中既有读取，也有写入。

tcp 的两个端点，一个是客户端，一个是服务端。

- 客户端：对应的对象，Socket。
- 服务端：对应的对象，ServerSocket。

3. UDP 传输

- 只要是网络传输，必须有 Socket。
- 数据一定要封装到数据包中，数据包中包括目的地址、端口、数据等信息。

直接操作 UDP 不可能，对于 Java 语言应该将 UDP 封装成对象，易于我们的使用，这个对象就是 DatagramSocket，即封装了 UDP 传输协议的 socket 对象。

因为数据包中包含的信息较多，为了方便操作这些信息，也一样会将其封装成对象。这个数据包对象就是 DatagramPacket。通过这个对象中的方法，就可以获取到数据包中的各种信息。

DatagramSocket 具备发送和接受功能，在进行 UDP 传输时，需要明确一个是发送端，一个是接收端。

7.7 课后习题

一、填空题

1. Socket 技术是构建在________协议之上的。
2. Datagrams 技术是构建在________协议之上的。
3. ServerSocket.accept() 返回________对象，是服务器与客户端相连。

二、选择题

1. URL 地址由________组成。

A. 文件名和主机名　　B. 主机名和端口号
C. 协议名和资源名　　D. IP 地址和主机名

2. IP 地址封装类是________。

A. InetAddress 类　　B. Socket 类
C. URL 类　　D. ServerSocket 类

3. InetAddress 类中获得主机名的方法是________。

A. getFile()　　B. getHostName()
C. getPath()　　D. getHostAddress()

4. Java 中面向无连接的数据报通信的类有________。

A. DatagramPacket 类　　B. DatagramSocket 类
C. DatagramPacket 类和 DatagramSocket 类　　D. Socket 类

三、简答题

1. 什么是 URL，它由哪几部分组成?
2. URLConnection 类与 URL 类有什么区别?
3. 客户机和服务器模式有什么特点? Socket 类和 ServerSocket 类的区别是什么?

7.8 上机实训

实训一、通过 URL 获取 www.baidu.com 的信息（端口号、IP、协议名、主机名、路径等）。

实训二、通过示例代码，编写服务器 - 客户端一对一聊天程序。

实训三、用 UDP 方式发送数据报给所有机房电脑信息（先给其他主机写好接收程序）。

实训四、编写文件下载服务器，功能如下。

- 当客户端连接服务器后，列出服务器上所有文件及文件夹。
- 客户端可输入文件或文件夹的名称进行文件下载。
- 如文件存在，则下载到指定文件夹中，如不存在，则给出对应提示。

PART08

知识领域8 Java数据库编程

知识目标

1. 理解数据库访问技术。
2. 掌握JDBC连接方式。
3. 掌握访问数据库、处理结果集的方法。
4. 了解数据库连接池。

■ 能力目标

1. 会使用Connection、Statement、ResultSet以及PreparedStatement类。
2. 熟练编写DBManger.Java工具类来连接数据库。
3. 熟练使用JDBC对数据库进行CRUD操作。
4. 能将查询的结果用Jtalbe或Jtextfield表示。

■ 素质目标

1. 能够阅读科技文档和撰写分析文档。
2. 能够查阅JDK API。
3. 增强学生团队协作能力。

8.1 应用场景

大多数软件系统都需要处理非常庞大的数据，这些数据并不是使用数据或集合就能解决的，这时就需要借助数据库系统。数据库系统由数据库、数据库管理系统、应用系统和数据库管理员组成。数据库管理系统简称 DBMS。目前有许多 DBMS 产品，如 DB2、Oracle、Microsoft SQL Server、Sybase、Informix、MySQL 等，Java 程序需要访问这些数据库，并对数据进行处理。

例如现在要为校园图书馆开发一个图书管理系统 Books Management System（BMS），学校使用的是 MySQL 数据库存储图书信息，图书表如下（其中编号为自增列）。

编号（bNo）	书名（title）	作者（author）	类型（bType）	数量（number）

使用 Java 编程，配合 Java 提供的 GUI 库和 JDBC 技术完成图书的增删查改操作。

8.2 相关知识

8.2.1 数据库访问技术简介

数据库中的数据存放在数据库文件中，我们要从数据库文件中获取数据，先要连接并登录到存放数据库的服务器。一般来说，访问数据库中的数据有以下两种方式。

一是通过 DBMS（Data Base Management System，数据库管理系统）提供的数据库操作工具来访问，如通过 SQL Server 2000 的查询设计器来提交查询，或者通过 SQL Server 2000 的企业管理工具来访问。这种方式比较适合 DBA 对数据库进行管理。

二是通过 API（Application Programming Interface，应用编程接口）来访问数据库，这种方式适合在应用程序中访问数据库。

在数据库发展的初期，各个开发商为自己的数据库设计了各自不同的 DBMS，因此不同类型的数据库之间交换数据非常困难。为了解决这个问题，Microsoft 提出了 ODBC（Open Data Base Connectivity，开放数据库互连）技术，试图建立一种统一的应用程序访问数据库接口，使开发人员无须了解程序内部结构就可以访问数据库。

JDBC（Java Data Base Connectivity,Java 数据库连接）是一种用于执行 SQL 语句的 Java API，可以为多种关系数据库提供统一访问，它由一组用 Java 语言编写的类和接口组成。JDBC 提供了一种基准，据此可以构建更高级的工具和接口，使数据库开发人员能够编写数据库应用程序。JDBC 并不能直接访问数据库，需要借助于数据库厂商提供的 JDBC 驱动程序。

JDBC 中常用的类和接口可用于我们编程开发，利用这些类和接口可以方便地进行数据访问和处理。这些类和接口都位于 Java.sql 包中。

8.2.2 JDBC 连接数据库

1. jdbc-odbc 桥连接

（1）ODBC 简介。

ODBC 是微软公司开放服务结构中有关数据库的一个组成部分，它建立了一组规范，并提供了一组对数据库访问的标准 API。应用程序可以使用所提供的 API 来访问任何提供了 ODBC 驱动程序的数据库。ODBC 规范为应用程序提供了一套高层调用接口规范和基于动态链接的运行支持环境。ODBC 已经成为一种标准，目前所有的关系数据库都提供了 ODBC 驱动程序，使用 ODBC 开发的应用程序具有很好的适应性和可移植性，并且具有同时访问多种数据库系统的能力。这使得 ODBC 的应用非常广泛，基本可用于所有的关系数据库。

要使用 ODBC，先要了解以下概念：ODBC 驱动管理器、ODBC 驱动程序、数据源。它们都是 ODBC 的组件。ODBC 组件之间的关系如图 8-1 所示。

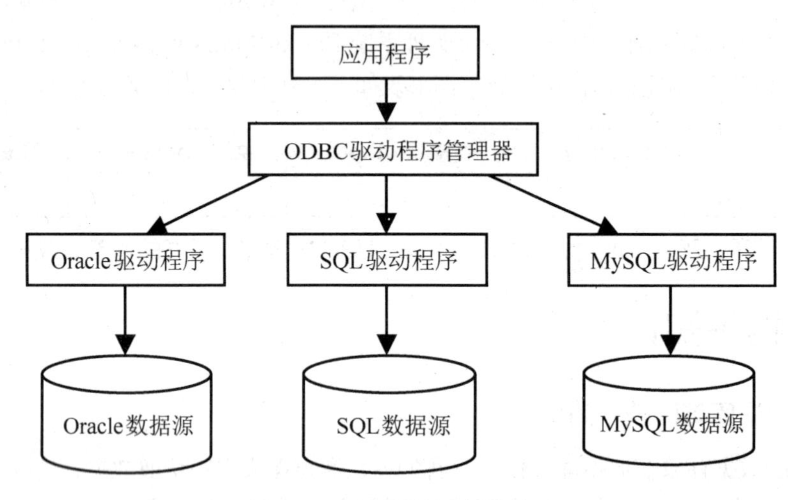

图8-1　ODBC组件之间的关系

① ODBC 驱动程序管理器。

应用程序不是直接调用 ODBC 驱动程序，而是先调用 ODBC 驱动程序管理器提供的 API，而 ODBC 驱动程序管理器再调用相应的 ODBC 驱动程序。这种间接的调用使得不管是连接到什么数据库都可以按照一定的方式来调用。

ODBC 驱动程序管理器负责将适当的 ODBC 驱动程序加载到内存中，并将应用程序的请求发给正确的 ODBC 驱动程序。ODBC 驱动程序管理器代表应用程序加载 ODBC 数据库驱动程序的动态链接库（ODBC32.dll）。该 DLL（Dynamic Link Librarry）对应用程序是透明的。

② ODBC 驱动程序。

ODBC 驱动程序处理从 ODBC 驱动程序管理器发送过来的函数调用，它负责将 SQL 请求发给相应的 DBMS，并将结果返回给 ODBC 驱动程序管理器。每个遵循 ODBC 的数据库应该提供自己的 ODBC 驱动程序，不同数据源的 ODBC 驱动程序不能混用。

③ 数据源。

数据源是数据、访问该数据所需要的信息和该数据源位置的特定集合，其中的数据源位置可用数据源名称描述。例如，数据源可以是通过网络在 Microsoft SQL Server 上运行的远程数据库，也可以是本地目录中的 Microsoft Access 数据库。用户只需用定义好的数据源名称访问数据库，而无须知道其他细节。通过应用程序，可以访问任何具有 ODBC 驱动程序的数据源。如 SQL Server、Oracle、Access 等。

（2）JDBC 简介。

之前我们介绍的数据库访问技术都是 Microsoft 公司提出的，主要用于 Windows 平台上

Microsoft 开发环境下的数据库连接和操作。而 JDBC（Java Data Base Connectivity，Java 数据库连接）是 Java 语言中用来规范客户端程序如何访问数据库的应用程序接口。

JDBC 是一种用于执行 SQL 语句的 Java API，可以为多种关系数据库提供统一访问，它由一组用 Java 语言编写的类和接口组成。JDBC 保留 ODBC 的基本设计功能，在 Web 和 Internet 应用程序中的作用与 ODBC 在 Windows 系列平台应用程序中的作用类似，而且还具有对硬件平台、操作系统异构性的支持。

JDBC 总体结构类似于 ODBC，如图 8-2 所示。

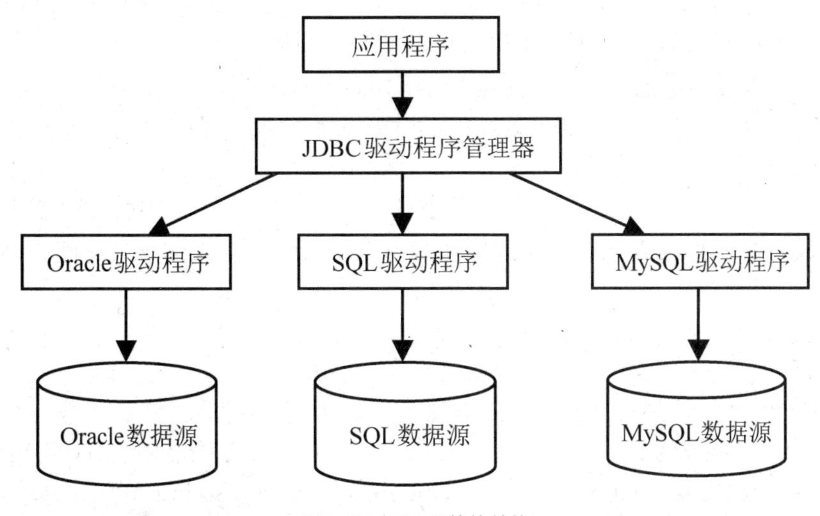

图8-2　JDBC整体结构

JDBC 应用程序负责用户与用户接口之间的交互操作。驱动程序管理器为应用程序加载和调用数据库驱动程序。数据库驱动程序执行 JDBC 对象方法的调用，发 SQL 请求给指定的数据源，并将结果返回给应用程序。驱动程序也负责与任何访问数据源的必要软件层进行交互。

（3）JDBC-ODBC 桥。

利用 ODBC 驱动程序提供 JDBC 访问。通过 JDBC-ODBC 桥，开发人员可以使用 JDBC 来存取 ODBC 数据源。不足的是，他需要在客户端安装 ODBC 驱动程序，换句话说，必须安装 Microsoft Windows 的某个版本。使用这一类型你需要牺牲 JDBC 的平台独立性。另外，ODBC 驱动程序还需要具有客户端的控制权限。另外，JDBC 驱动程序还有本地 API 驱动、网络协议驱动以及本地协议驱动 3 种方式。

2. 纯 Java 方式连接

即本地协议驱动，这种方式直接把 JDBC 调用转换为符合相关数据库系统规范的请求。由这种驱动写的应用可以直接和数据库服务器通信。这种类型的驱动完全由 Java 实现。

8.2.3　执行 SQL 操作

访问数据库就是从数据库中提取数据或向数据库中保存数据，JDBC 抽象了数据库进行交互的过程。首先在 Java 程序中要使用 import java sql.* 语句导入 Java.sql 包。当 Java 程序需要使用 JDBC 时，创建一个 Java.sql.connection 对象指向数据库。其次，要创建一个最基本的类 Java.sql.statement，用于执行数据库操作。

用 JDBC 来实现访问数据库的具体操作步骤如下。

（1）加载连接数据库的驱动程序。

在指定数据库建立连接前，JDBC 都会加载相应的驱动程序，每一种数据库的驱动程序完成 Java.sql.Driver 接口。加载驱动程序的一种最简单的方法是使用 Class.forName() 方法显示加载一个驱动程序（这里使用的是 MySQL 数据库，所以加载的是 MySQL 驱动），由驱动程序负责向 DriverManager 登记注册并在与数据库相连接时，DriverManager 将使用此驱动程序。如：

```
Class.forName("com.mysql.jdbc.Driver");
```

（2）建立连接。

采用 DriverManager 类中的 getConnection 方法实现与 url 所指定的数据源建立连接并返回一个 Connection 类的对象，以后对这个数据源的操作都是基于该 Connection 类对象。

```
String url = "jdbc:mysql://localhost:3306/bms";
    connection con = DriverManager.getConnection(url, "root","123456");
```

（3）查询数据库。

对于需要返回结果集的 SELECT 语句，使用 executeQuery() 方法，该方法只有一个字符串参数，用来存放 SELECT 语句，查询成功则以 Resultset 对象的形式返回查询结果。使用 Statement 对象来执行。

```
    Statement stmt = con.createStatement();
    String sql = "SELECT  * FROM  book";
    ResultSet rs = stmt. executeQuery(sql);
```

如果对数据库系统发送 INSERT、UPDATE 和 DELETE 等不需要返回查询结果的 SQL 语句，则采用 executeUpdate() 方法。executeUpdate() 方法接受 String 类型的 SQL 语句为参数，返回类型为 int，表示数据库表受到 INSERT、UPDATE 和 DELETE 语句影响的数据行数。如果返回值为 0，表示 SQL 语句不返回任何数据。

```
Statement stmt = con.createStatement();
String sql = "DELETE FROM book WHERE bNo = 2";
int ret = stmt. executeUpdate(sql);
```

（4）处理结果集。

对 ResultSet 对象进行处理后，可以将查询结果显示给用户。ResultSet 对象包括一个包含所有查询结果的表。最初，游标位于结果集的第一行的前面，可以用 ResultSet.next 方法使指针下移一行对结果逐行处理，并用 ResultSet 类的 get 方法将数据库各个字段名类型转换为 Java 类型。

```
while(rs.next()){//当ResultSet对象rs指到最后一个的下一个时，返回false
    int bNo = rs.getInt("bNo");
    String title = rs.getString("title");
    String author = rs.getString("author");
    String bType = rs.getString("bType");
    int number = rs.getInt("number");
}
```

（5）关闭查询语句及数据库连接。

在程序结束前应依次关闭 Statement 对象和 Connection 对象，使用 close() 方法。

```
st.close();
    conn.close();
```

8.3 任务实施

任务 编写程序实现对图书信息表的增删改查操作

1. 任务需求

现有图书信息表 bookinfo（序号，图书编号，图书名称，作者，出版社，单价，出版日期，ISBN，库存数量）。写程序实现对该表的增删改查操作。

2. 任务分析

（1）连接 MySQL 使用的数据库连接包是：mysql-connector-Java-3.1.14-bin.jar。

（2）由于图书信息较多，我们编写 Book.java 来存放信息。

（3）因为每次执行 SQL 操作都要连接数据库，所以我们编写 DBManager.java 来统一管理数据库的连接和关闭，然后把数据库连接操作统一写进 BookDAO.java。

本任务类图如图 8-3 所示。

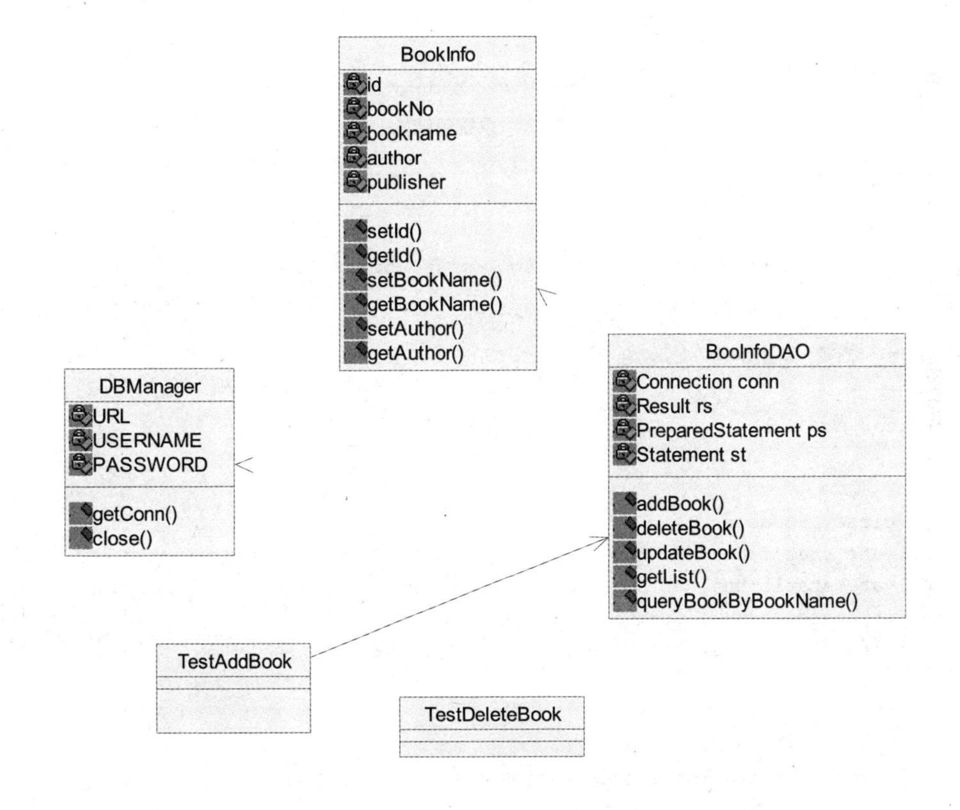

图8-3 类图

DBManager.java，实现数据库的连接和关闭。

BookInfo.java，存储图书的基本信息。

BookDAO.java，完成对数据库表的增删改查等基本操作。

系统包结构如图 8-4 所示。

```
▲ com.daiinfo.seniorjava.ken8.implement.bean
    ▷ BookInfo.java
▲ com.daiinfo.seniorjava.ken8.implement.dao
    ▷ BookInfoDAO.java
▲ com.daiinfo.seniorjava.ken8.implement.test
    ▷ SimpleFrame.java
    ▷ TestAddBook.java
    ▷ TestDeleteBook.java
    ▷ TestList.java
    ▷ TestQueryBybookname.java
    ▷ TestShowOnTable.java
    ▷ TestShowOnTextArea.java
    ▷ TestUpdateBook.java
▲ com.daiinfo.seniorjava.ken8.implement.utils
    ▷ DBConnPool.java
    ▲ DBManager.java
        ▷ DBManager
      bookinfo.sql
```

图8-4　系统包结构

3. 任务实现

DBManager.java

```java
package com.daiinfo.seniorjava.ken8.implement.utils;
import java.sql.*;
/**
 * 数据库的连接和关闭等管理类
 *
 * @author 戴远泉
 *
 */
public class DBManager {
    private static final String DRIVER = "com.mysql.jdbc.Driver";
    private static final String URL = "jdbc:mysql://localhost:3306/test";
    private static final String USRENAME = "root";
    private static final String PASSWORD = "123456";
    static {
        try {
            Class.forName(DRIVER);
        } catch (ClassNotFoundException e) {
            e.printStackTrace();
        }
    }
    public static void main(String[] args) {
        System.out.print(getConn());
    }
    public static Connection getConn() {
        Connection conn = null;
        try {
```

```
            conn = DriverManager.getConnection(URL, USRENAME, PASSWORD);
        } catch (SQLException e) {
            e.printStackTrace();
        }
        return conn;
    }
    public static void close(Connection conn, Statement st, ResultSet rs) {
        try {
            if (rs != null) {
                rs.close();
            }
            if (st != null) {
                st.close();
            }
            if (conn != null) {
                conn.close();
            }
        } catch (Exception e) {
            // TODO Auto-generated catch block
            e.printStackTrace();
        }
    }
}
```

BookInfo.java

```
package com.daiinfo.seniorjava.ken8.implement.bean;
import java.io.Serializable;
/**
 * 图书信息类，实体类，存放图书的基本信息及基本操作
 *
 * @author 戴远泉
 * @version 1.0
 */
public class BookInfo implements Serializable {
    int id;
    String bookNo;
    String bookname;
    String author;
    String publisher;
    double price;
    String publishtime;
    String ISBN;
    int amount;
    public BookInfo() {
    }
    public BookInfo(String bookNo, String bookname, String author) {
        this.bookNo = bookNo;
        this.bookname = bookname;
        this.author = author;
    }
    public int getId() {
```

```
        return id;
    }
    public void setId(int id) {
        this.id = id;
    }
    public String getBookNo() {
        return bookNo;
    }
    public void setBookNo(String bookNo) {
        this.bookNo = bookNo;
    }
    public String getBookname() {
        return bookname;
    }
    public void setBookname(String bookname) {
        this.bookname = bookname;
    }
    public String getAuthor() {
        return author;
    }
    public void setAuthor(String author) {
        this.author = author;
    }
    public String getPublisher() {
        return publisher;
    }
    public void setPublisher(String publisher) {
        this.publisher = publisher;
    }
    public double getPrice() {
        return price;
    }
    public void setPrice(double d) {
        this.price = d;
    }
    public String getPublishtime() {
        return publishtime;
    }
    public void setPublishtime(String publishtime) {
        this.publishtime = publishtime;
    }
    public String getISBN() {
        return ISBN;
    }
    public void setISBN(String iSBN) {
        ISBN = iSBN;
    }
    public int getAmount() {
        return amount;
    }
    public void setAmount(int amount) {
```

```
        this.amount = amount;
    }
    @Override
    public String toString() {
        return bookNo + bookname + author;
    }
}
```

BookInfoDAO.java

```
package com.daiinfo.seniorjava.ken8.implement.dao;
import java.sql.Connection;
import java.sql.PreparedStatement;
import java.sql.ResultSet;
import java.sql.SQLException;
import java.sql.Statement;
import java.util.ArrayList;
import java.util.List;
import com.daiinfo.seniorjava.ken8.implement.bean.BookInfo;
import com.daiinfo.seniorjava.ken8.implement.utils.DBManager;
/**
 * 数据访问类，封装对数据库表的增删改查操作
 *
 * @author 戴远泉
 * @version 1.0
 */
public class BookInfoDAO {
    private Connection conn;
    private ResultSet rs;
    private PreparedStatement ps;
    private Statement st;
    /**
     * 添加图书
     *
     * @param book
     * @return 布尔类型值
     */
    public boolean addBook(BookInfo book) {
        boolean flag = false;
        conn = DBManager.getConn();
        String sqlString = "insert into bookinfo
        (bookNo,bookname,author,publisher,price,publishtime,ISBN,amount) values(?,?,?,?,?,?,?,?)";
        try {
            ps = conn.prepareStatement(sqlString);
            ps.setString(1, book.getBookNo());
            ps.setString(2, book.getBookname());
            ps.setString(3, book.getAuthor());
            ps.setString(4, book.getPublisher());
            ps.setDouble(5, book.getPrice());
            ps.setString(6, book.getPublishtime());
            ps.setString(7, book.getISBN());
            ps.setInt(8, book.getAmount());
```

```
            flag = ps.executeUpdate() != 0 ? true : false;
        } catch (Exception e) {
            e.printStackTrace();
        }
        return flag;
    }
    /**
     * 根据图书名查询
     *
     * @param bname
     * @return 图书列表
     */
    public List<BookInfo> queryByBookName(String bname) {
        List<BookInfo> books = new ArrayList<BookInfo>();
        conn = DBManager.getConn();
        String sqlString = "select * from bookinfo where bookname like ?";
        try {
            ps = conn.prepareStatement(sqlString);
            // bname = "%" + bname + "%";
            ps.setString(1, bname);
            ResultSet rs = ps.executeQuery();
            while (rs.next()) {
                int bookId = rs.getInt("ID");
                String bookNo = rs.getString("bookNo");
                String bookname = rs.getString("bookname");
                String author = rs.getString("author");
                String publisher = rs.getString("publisher");
                String publishtime = rs.getString("publishtime");
                double price = rs.getFloat("price");
                String ISBN = rs.getString("ISBN");
                int amount = rs.getInt("amount");
                BookInfo book = new BookInfo(bookNo, bookname, author);
                book.setId(bookId);
                book.setPublisher(publisher);
                book.setPublishtime(publishtime);
                book.setISBN(ISBN);
                book.setPrice(price);
                book.setAmount(amount);
                books.add(book);
            }
        } catch (SQLException e) {
            // TODO Auto-generated catch block
            e.printStackTrace();
        }
        return books;
    }
    /**
     * 查询所有的数据列表
     *
     * @return 图书结果集
     */
```

```
    public List<BookInfo> getList() {
        List<BookInfo> books = new ArrayList<BookInfo>();
        conn = DBManager.getConn();
        String sql = "select * from bookinfo";
        try {
            /*
             * st = conn.createStatement(); rs = st.executeQuery(sql);
             */
            ps = conn.prepareStatement(sql);
            rs = ps.executeQuery();
            while (rs.next()) {
                BookInfo book = new BookInfo();
                book.setId(rs.getInt(1));
                book.setBookNo(rs.getString(2));
                book.setBookname(rs.getString(3));
                book.setAuthor(rs.getString(4));
                book.setPublisher(rs.getString(5));
                book.setPrice(rs.getDouble(6));
                book.setPublishtime(rs.getString(7));
                book.setISBN(rs.getString(8));
                book.setAmount(rs.getInt(9));
                books.add(book);
            }
        } catch (SQLException e) {
            // TODO Auto-generated catch block
            e.printStackTrace();
        }
        return books;
    }
    /**
     * 根据图书的ID删除图书
     *
     * @param id
     * @return
     */
    public boolean delBookById(int id) {
        boolean flag = false;
        conn = DBManager.getConn();
        String sql = "delete from bookinfo where ID=?";
        try {
            ps = conn.prepareStatement(sql);
            ps.setInt(1, id);
            if (ps.execute()) {
                flag = true;
            } else {
                flag = false;
            }
        } catch (SQLException e) {
            // TODO Auto-generated catch block
            e.printStackTrace();
        }
```

```
            return flag;
        }
        /**
         * 根据图书进行修改其信息
         *
         * @param book
         * @return
         */
        public boolean update(BookInfo book) {
            boolean flag = false;
            conn = DBManager.getConn();
            String sql = "update bookinfo set bookNo = ?,bookname = ?,author=?,publisher
=?,price=?,publishtime=?,ISBN=?,amount=? where ID=?";
            try {
                ps = conn.prepareStatement(sql);
                ps.setString(1, book.getBookNo());
                ps.setString(2, book.getBookname());
                ps.setString(3, book.getAuthor());
                ps.setString(4, book.getPublisher());
                ps.setDouble(5, book.getPrice());
                ps.setString(6, book.getPublishtime());
                ps.setString(7, book.getISBN());
                ps.setInt(8, book.getAmount());
                ps.setInt(9, book.getId());
                flag = ps.executeUpdate() > 0 ? true : false;
            } catch (Exception e) {
                e.printStackTrace();
            }
            return flag;
        }
    }
```

（1）向数据库表中插入数据。

TestAddBook.java

```
    package com.daiinfo.seniorjava.ken8.implement.test;
    import com.daiinfo.seniorjava.ken8.implement.bean.BookInfo;
    import com.daiinfo.seniorjava.ken8.implement.dao.BookInfoDAO;
    public class TestAddBook {
        public static void main(String[] args) {
            // TODO Auto-generated method stub
            BookInfoDAO bookInfoDAO = new BookInfoDAO();
            BookInfo book = new BookInfo(null, null, null);
            book.setBookNo("08803");
            book.setBookname("java高级程序设计");
            book.setAuthor("戴远泉");
            book.setPublisher("高等教育出版社");
            book.setPublishtime("20170809");
            book.setAmount(4);
            book.setPrice(45.05);
            book.setISBN("98745372123");
            if(bookInfoDAO.addBook(book))
```

```
            System.out.println("添加成功！");
        else {
            System.out.println("添加失败！");
        }
    }
}
```

执行结果如图 8-5 所示。

添加成功

[表] bookinfo @test (localhost)

文件(F) 编辑(E) 查看(V) 窗口(W)

导入向导(I) 导出向导(X) 筛检向导 网格视图 表单视图 备注 十六进位 图像 升幂排序

ID	bookNo	bookname	author	publisher
2	0204	Java高级程序设计	戴远泉	电子工业出版社
3	0203	操作系统	戴远泉	清华大学出版社
4	0205	数据结构	戴远泉	(Null)
8	08803	Java高级程序设计	戴远泉	高等教育出版社

图8-5 添加数据

（2）删除图书数据。

TestDeleteBook.java

```
package com.daiinfo.seniorjava.ken8.implement.test;
import java.util.ArrayList;
import java.util.List;
import com.daiinfo.seniorjava.ken8.implement.bean.BookInfo;
import com.daiinfo.seniorjava.ken8.implement.dao.BookInfoDAO;
public class TestDeleteBook {
    public static void main(String[] args) {
        BookInfo book = new BookInfo();
        book.setBookname("java程序设计");
        System.out.println("________________________");
        BookInfoDAO bookInfoDAO = new BookInfoDAO();
        List<BookInfo> books = bookInfoDAO.queryByBookName(book.getBookname());
        if (books.size() >= 1) {
            int bookid = books.get(0).getId();
            if (!bookInfoDAO.delBookById(bookid)) {
                System.out.println("删除成功！");
            } else {
                System.out.println("删除失败！");
            }
        }
    }
}
```

执行结果如图 8-6 所示。

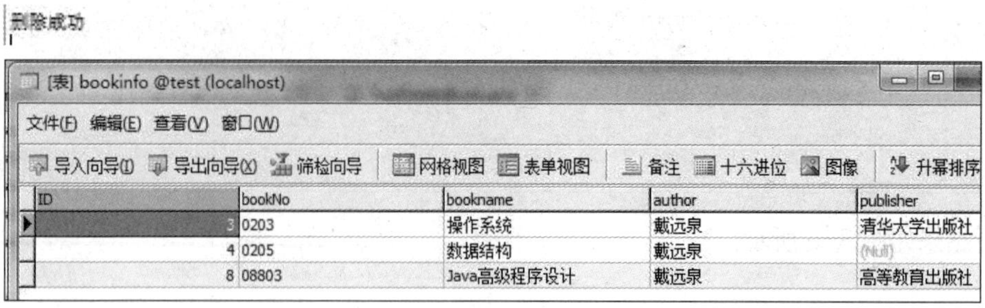

删除成功

[表] bookinfo @test (localhost)

文件(F) 编辑(E) 查看(V) 窗口(W)

导入向导(I) 导出向导(X) 筛检向导 网格视图 表单视图 备注 十六进位 图像 升幂排序

ID	bookNo	bookname	author	publisher
3	0203	操作系统	戴远泉	清华大学出版社
4	0205	数据结构	戴远泉	(Null)
8	08803	Java高级程序设计	戴远泉	高等教育出版社

图8-6　删除数据

（3）修改相应图书信息。

TestUpdateBook.java

```
package com.daiinfo.seniorjava.ken8.implement.test;
import com.daiinfo.seniorjava.ken8.implement.bean.BookInfo;
import com.daiinfo.seniorjava.ken8.implement.dao.BookInfoDAO;
public class TestUpdateBook {
    public static void main(String[] args) {
        BookInfoDAO bdao = new BookInfoDAO();
        BookInfo book = new BookInfo("0203", "操作系统", "戴远泉");
        java.util.List<BookInfo> books = new java.util.ArrayList<BookInfo>();
        books = bdao.queryByBookName(book.getBookname());
        for (BookInfo bookInfo : books) {
            System.out.println(bookInfo.getId() + "\t" + bookInfo.getBookNo()
                    + "\t" + bookInfo.getBookname() + "\t"
                    + bookInfo.getAuthor());
        }
        book.setId(books.get(0).getId());
        book.setBookname("操作系统原理");
        book.setAuthor("王欣");
        book.setPublisher("华中科技大学出版社");
        book.setPublishtime("10170801");
        if (bdao.update(book)) {
            System.out.println("修改成功! ");
        } else {
            System.out.println("修改失败! ");
        }
    }
}
```

执行结果如图 8-7 所示。

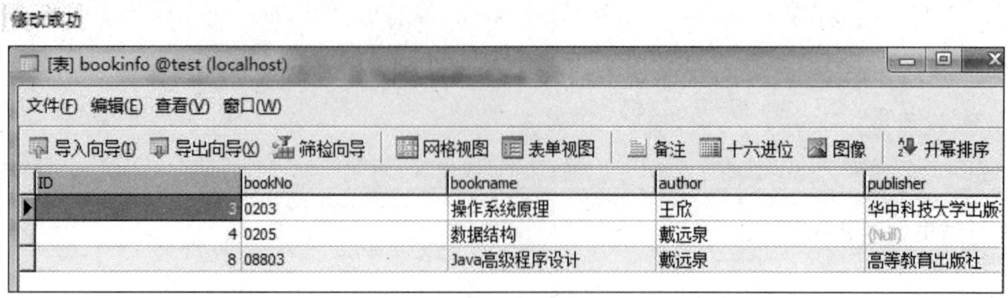

修改成功

[表] bookinfo @test (localhost)

文件(F) 编辑(E) 查看(V) 窗口(W)

导入向导(I) 导出向导(X) 筛检向导 网格视图 表单视图 备注 十六进位 图像 升幂排序

ID	bookNo	bookname	author	publisher
3	0203	操作系统原理	王欣	华中科技大学出版
4	0205	数据结构	戴远泉	(Null)
8	08803	Java高级程序设计	戴远泉	高等教育出版社

图8-7　修改图书信息

（4）查询图书信息。

```
package com.daiinfo.seniorjava.ken8.implement.test;
import java.sql.ResultSet;
import java.sql.SQLException;
import java.util.List;
import com.daiinfo.seniorjava.ken8.implement.bean.BookInfo;
import com.daiinfo.seniorjava.ken8.implement.dao.BookInfoDAO;
public class TestQueryBybookname {
    public static void main(String[] args) {
        BookInfoDAO bookInfoDAO = new BookInfoDAO();
        List<BookInfo> books = bookInfoDAO.queryByBookName("java程序设计");
        for (BookInfo bookInfo : books) {
            System.out.println(bookInfo.getId() + "\t" + bookInfo.getBookNo()
                    + "\t" + bookInfo.getBookname() + "\t"
                    + bookInfo.getAuthor());
        }
    }
}
```

运行结果如图 8-8 所示。

```
Markers  Properties  Servers  Data Source Explorer  Snippets  Console  Search
<terminated> TestQueryBybookname [Java Application] C:\Java\jdk1.7.0_67\bin\javaw.exe (2017年8月24日 下午5:39:39)
9	0908	Java程序设计	戴远泉
10	0902	Java程序设计	张和
11	0906	Java程序设计	刘珂
```

图8-8　运行结果

（5）显示数据。

前面 4 步已经完成了基本的增删查改操作。现在将结合所学习过的 Java GUI 的知识，把数据分别显示到一个 JTalbe 和一个 JTextfield 中。

① 在 JTable 中显示数据。

TestShowOnTable.java

```
package com.daiinfo.seniorjava.ken8.implement.test;
s
import java.awt.*;
import java.util.List;
import javax.swing.*;
import javax.swing.table.*;
import com.daiinfo.seniorjava.ken8.implement.dao.BookInfoDAO;
import com.daiinfo.seniorjava.ken8.implement.bean.BookInfo;
public class TestShowOnTable extends JFrame {
    // 定义表格
    JTable table;
    // 定义滚动条面板（用以使表格可以滚动）
    JScrollPane scrollPane;
    // 定义数据模型类的对象（用以保存数据），
    DefaultTableModel tableModel;
    public TestShowOnTable() {
        super();
        setTitle("图书表");
```

```
        scrollPane = new JScrollPane();
        // 定义表格列名数组
        final String[] columnNames = { "序号", "图书编号", "书名", "作者", "出版社", "单价",
                "出版日期", "ISBN", "库存数量" };
        // 表格数据数组
        BookInfoDAO bdao = new BookInfoDAO();
        List<BookInfo> books = bdao.getList();
        String tableValues[][] = new String[books.size()][9];
        for (int i = 0; i < books.size(); i++) {
            BookInfo book = books.get(i);
            tableValues[i][0] = Integer.toString(book.getId());
            tableValues[i][1] = book.getBookNo();
            tableValues[i][2] = book.getBookname();
            tableValues[i][3] = book.getAuthor();
            tableValues[i][4] = book.getPublisher();
            tableValues[i][5] = Double.toString(book.getPrice());
            tableValues[i][6] = book.getPublishtime();
            tableValues[i][7] = book.getISBN();
            tableValues[i][8] = Integer.toString(book.getAmount());
        }
        // 创建指定表格列名和表格数据的表格模型类的对象
        tableModel = new DefaultTableModel(tableValues, columnNames);
        // 创建指定表格模型的表格
        table = new JTable(tableModel);
        // 设置 RowSorter（RowSorter 用于提供对 JTable 的排序和过滤）
        table.setRowSorter(new TableRowSorter<DefaultTableModel>(tableModel));
        scrollPane.setViewportView(table);
        getContentPane().add(scrollPane, BorderLayout.CENTER);
        table.setEnabled(false);// 禁止修改
        setBounds(300, 200, 400, 300);
        setVisible(true);
        setDefaultCloseOperation(JFrame.EXIT_ON_CLOSE);
    }
    public static void main(String args[]) {
        new TestShowOnTable();
    }
}
```

运行结果如图 8-9 所示。

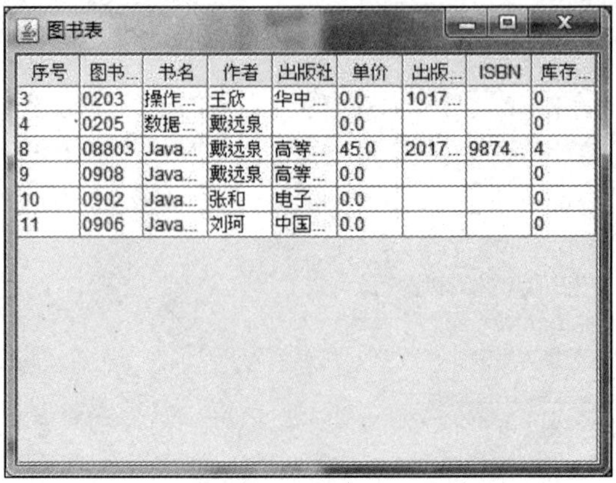

图书表

序号	图书...	书名	作者	出版社	单价	出版...	ISBN	库存...
3	0203	操作...	王欣	华中...	0.0	1017...		0
4	0205	数据...	戴远泉		0.0			0
8	08803	Java...	戴远泉	高等...	45.0	2017...	9874...	4
9	0908	Java...	戴远泉	高等...	0.0			0
10	0902	Java...	张和	电子...	0.0			0
11	0906	Java...	刘珂	中国...	0.0			0

图8-9　运行结果

② 在 JTextfield 中显示数据。

TestShowOnTextArea.java

```
package com.daiinfo.seniorjava.ken8.implement.test;
import java.awt.*;
import javax.swing.*;
import java.sql.Connection;
import java.sql.DatabaseMetaData;
import java.sql.ResultSet;
import java.sql.SQLException;
import java.util.List;
import com.daiinfo.seniorjava.ken8.implement.dao.BookInfoDAO;
import com.daiinfo.seniorjava.ken8.implement.bean.BookInfo;
public class TestShowOnTextArea {
    JFrame jf;
    JPanel jpanel;
    JTextArea jta = null;
    JScrollPane jscrollPane;
    BookInfoDAO bdao = new BookInfoDAO();
    Connection conn;
    private String m_TableName;
    public TestShowOnTextArea() {
        jf = new JFrame("显示结果");
        Container contentPane = jf.getContentPane();
        jta = new JTextArea(10, 15);
        jta.setEditable(false);
        jscrollPane = new JScrollPane(jta);
        contentPane.add(jscrollPane);
        jta.append("序号\t图书编号\t书名\t作者\t出版社\t单价\t出版日期\tISBN\t库存数量\n");
        List<BookInfo> books = bdao.getList();
        for (BookInfo book : books) {
            jta.append(Integer.toString(book.getId()));
            jta.append("\t" + book.getBookNo());
            jta.append("\t" + book.getBookname());
            jta.append("\t" + book.getAuthor());
            jta.append("\t" + book.getPublisher());
            jta.append("\t" + book.getPrice());
            jta.append("\t" + book.getPublishtime());
            jta.append("\t" + book.getISBN());
            jta.append("\t" + book.getAmount());
            jta.append("\n");
        }
        jf.setSize(500, 400);
        jf.setLocation(400, 200);
        jf.setVisible(true);
        jf.setDefaultCloseOperation(JFrame.EXIT_ON_CLOSE);
    }
    public static void main(String[] args) {
        new TestShowOnTextArea();
    }
}
```

运行结果如图 8-10 所示。

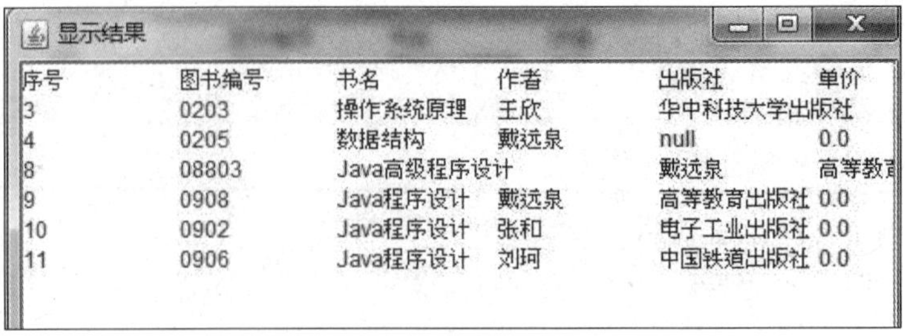

显示结果

序号	图书编号	书名	作者	出版社	单价
3	0203	操作系统原理	王欣	华中科技大学出版社	
4	0205	数据结构	戴远泉	null	0.0
8	08803	Java高级程序设计		戴远泉	高等教育
9	0908	Java程序设计	戴远泉	高等教育出版社	0.0
10	0902	Java程序设计	张和	电子工业出版社	0.0
11	0906	Java程序设计	刘珂	中国铁道出版社	0.0

图8-10 显示结果

8.4 拓展知识

数据库连接池负责分配、管理和释放数据库连接，它允许应用程序重复使用一个现有的数据库连接，而不是再重新建立一个；释放空闲时间超过最大空闲时间的数据库连接来避免因为没有释放数据库连接而引起的数据库连接遗漏。这项技术能明显提高对数据库操作的性能。

数据库连接池在初始化时将创建一定数量的数据库连接放到连接池中，这些数据库连接的数量是由最小数据库连接数制约的。无论这些数据库连接是否被使用，连接池都将一直保证至少拥有这么多的连接数量。连接池的最大数据库连接数量限定了这个连接池能占有的最大连接数，当应用程序向连接池请求的连接数超过最大连接数量时，这些请求将被加入到等待队列中。数据库连接池的最小连接数和最大连接数的设置要考虑到下列几个因素。

1. 最小连接数

是连接池一直保持的数据库连接，所以如果应用程序对数据库连接的使用量不大，将会有大量的数据库连接资源被浪费。

2. 最大连接数

是连接池能申请的最大连接数，如果数据库连接请求超过此数，后面的数据库连接请求将被加入到等待队列中，这会影响之后的数据库操作。

3. 最小连接数与最大连接数差距

最小连接数与最大连接数相差太大，那么最先的连接请求将会获利，之后超过最小连接数量的连接请求等价于建立一个新的数据库连接。不过，这些大于最小连接数的数据库连接在使用完不会马上被释放，它将被放到连接池中等待重复使用或是空闲超时后被释放。

下面我们编写 DBCPool.java 类来创建一个数据库连接池来获取连接，这里要用到 3 个 jar 包：commons-collections-3.1.jar、commons-dbcp-1.2.1.jar 以及 commons-pool-1.2.jar。

DBConnPool.java

```
package com.daiinfo.seniorjava.ken8.implement.utils;
import java.sql.*;
import org.apache.commons.dbcp.BasicDataSource;
public class DBConnPool {
    private static final String DRIVER = "com.mysql.jdbc.Driver";
    private static final String URL = "jdbc:mysql://localhost:3306/test";
    private static final String USRENAME = "root";
```

```
    private static final String PASSWORD = "123456";
    private static BasicDataSource bds;
    static {
        bds = new BasicDataSource();
        bds.setDriverClassName(DRIVER);// 设置驱动类
        bds.setUrl(URL);// 设置连接路径
        bds.setUsername(USRENAME);// 设置连接数据库管理员账号
        bds.setPassword(PASSWORD);// 设置连接数据库密码
        bds.setMinIdle(5);// 最小连接数
        bds.setMaxActive(20);// 最大连接数
    }
    // 获取连接的方法
    public static Connection getConn() {
        Connection conn = null;
        try {
            conn = bds.getConnection();
        } catch (SQLException e) {
            e.printStackTrace();
        }
        return conn;
    }
}
```

8.5 拓展训练

1. 任务需求

实现批量插入数据和批量删除数据。

2. 任务分析

当进行多条信息删除的时候，需要使用 Java 的事务处置机制对数据库进行删除，也就是说删除的时候如果选中的要删除的所有信息其中一条没有成功删除的话，那么就都不删除。

3. 任务实现

（1）将要批量删除的图书的 ID 拼接成使用“,”隔开的字符串，使用 BookInfoDAO 中的 deleteBookByID() 方法进行删除。

（2）在 BookInfoDAO.java 中增加方法 public boolean deleteBatch(String param)。

```
/**
     * 批量删除 将图书的ID组合成字符串，并用“,”隔开
     *
     * @param param
     * @return
     */
    public boolean deleteBatch(String param) {
        Boolean flag = false;
        conn = DBManager.getConn();
        String[] strings = param.split(",");
        try {
            conn.setAutoCommit(false);
            conn.commit();// 提交事务
```

```
            for (String s : strings) {
                int id = Integer.parseInt(s);
                delBookById(id);
            }
            flag = true;
        } catch (SQLException e) {
            try {
                conn.rollback(); // 进行事务回滚
            } catch (SQLException ex) {
                ex.printStackTrace();
            }
        }
        return flag;
    }
```

（3）TestDeleteBatch.java。

```
package com.daiinfo.seniorjava.ken8.implement.test;
import java.util.Iterator;
import com.daiinfo.seniorjava.ken8.implement.bean.BookInfo;
import com.daiinfo.seniorjava.ken8.implement.dao.BookInfoDAO;
public class TestDeleteBatch {
    public static void main(String[] args) {
        // TODO Auto-generated method stub
        String string = "15,17,18";
        String[] s = string.split(",");
        /*
         * for (String string2 : s) {
         * System.out.println(Integer.parseInt(string2)); }
         */
        BookInfoDAO bdao = new BookInfoDAO();
        BookInfo book = new BookInfo();
        if (bdao.deleteBatch(string)) {
            System.out.print("删除成功！");
        } else {
            System.out.print("删除失败！");
        }
    }
}
```

运行结果如图 8-11 所示。

ID	bookNo	bookname	author	publisher
11	0906	Java程序设计	刘珂	中国铁道出版社

图8-11　运行结果

8.6 课后小结

1. JDBC

JDBC（Java DataBase Connectivity，Java 数据库连接）是一种用于执行 SQL 语句的 JavaAPI，可以为多种关系数据库提供统一访问，它由一组用 Java 语言编写的类和接口组成。JDBC 为开发人员提供了一标准的 API，据此可以构建更高级的工具和接口，使数据库开发人员能够用纯 Java 的 API 编写数据库应用程序。

2. JDBC 连接数据库步骤

无论是普通 Java 程序、JavaWeb 应用程序、EJB 程序或是其他 Java 程序，使用 JDBC 操作数据库的步骤基本都是相对固定的。具体分为如下步骤。

① 注册驱动程序。即把驱动程序类加载到 Java 虚拟机中，使得驱动管理器 DriverManager 能够找到该驱动程序，一般通过 Class.forName() 进行加载。

② 获取数据库连接。Java.sql.Connection 接口代表一个数据库连接，它通过驱动管理器 DriverManager 来建立连接，并返回一个 Connection 接口的实现。同时需要指定连接 URL、用户名和密码。

③ 创建会话。JDBC 的会话 Statement 主要是用于向数据库发送 SQL 命令，并返回执行结果，由 Connection 生成。

④ 执行 SQL 语句。会话创建好后程序员就可以执行具体的 SQL 语句了。一般分为查询和修改两种。查询主要是 select 语句，使用 executeQuery() 方法，它返回查询后的结果集；修改包括对数据库记录的插入、修改、删除，使用 executeUpdate() 方法，它执行后返回的是影响到的记录数。

⑤ 处理结果集。对于查询语句，返回的是结果集 ResultSet，一般使用 ResultSet.next() 方法对结果集进行逐条处理。

⑥ 关闭连接。按照“结果集 => 会话 => 连接”的顺序关闭数据库连接。

8.7 课后习题

一、填空题

1. URL 定义了连接数据库时的协议、________和数据库标识。
2. JDBC API 提供的连接和操作数据库的类和接口位于________包和 Java.sql 包中。
3. ResultSet 对象的________方法表示将光标从当前位置移向下一行。

二、选择题

1. 下面是一组对 JDBC 的描述，正确的说法是________。

A. JDBC 是一个数据库管理系统

B. JDBC 是一个由类和接口组成的 API

C. JDBC 是一个驱动查询

D. JDBC 是一组命令

2. 创建数据库连接的目的是________。

A. 建立一条通往某个数据库的通道　　B. 加载数据库驱动程序

C. 清空数据库　　D. 为数据库增加记录

3. 要为数据库增加记录，应调用 Statement 对象的________方法。

A. addRecord() B. executeQuery() C. executeUpdate() D. executeAdd()

4. PreparedStatement 对象的________方法执行包含参数的动态 INSERT、UPDATE、DELETE 语句。

A. query() B. execute()

C. executeUpdate() D. executeQuery()

5. ResultSetMetaData 对象的________方法返回指定列序号的列名。

A. getColumnName() B. getColumnCount()

C. getColumnLabel() D. getColumnType()

6. 下列 SQL 语句中，哪一项可用 executeQuery() 方法发送到数据库________。

A. Update B. Delete

C. Select D. Insert

7. 下面哪个选项不是客户端的 Java 应用程序需要完成的工作________。

A. 与数据库的某个表建立连接 B. 与特定的数据库建立连接

C. 发送 SQL 语句，得到查询结果 D. 关闭与 JDBC 服务器的连接

三、简答题

1. 简述 Java.sql 包中主要类的作用。
2. 简单描述使用 JDBC 访问数据库的基本步骤。
3. 常用的数据库操作对象有哪些，这些对象分别用来做什么？
4. ResultSet 对象的作用是什么，该对象的常用方法有哪些？
5. 删除、增加、修改和查询记录的 SQL 语法是什么？其对应的 Statement 方法又是什么？

8.8 上机实训

实训一、参照书中代码，完成图书管理系统。要求：使用 MySQL 数据库自建图书表，字段为图书编号、书名、作者、分类、数量等（可自行添加），编写程序，用 JDBC 方式连接数据库完成以下功能。

- 新增一条图书信息，如果已经存在，则数量 +1；
- 删除一条图书信息，如果已经存在且数量不为 0，则数量 -1，有根据图书编号、书名两种方式删除；
- 修改一条图书信息，图书编号不可修改；
- 查询图书信息，有根据图书编号、书名、作者、分类 4 种方式查询；
- 把查询结果用 JTextFeild 或者 JTable 显示。

实训二、请结合反射和注解方面的知识，编写通用的程序，实现对任意对象的增删改查操作。

PART09

知识领域9

综合实训——基于C/S架构的餐饮管理系统的设计与实现

知识目标

1. 掌握餐饮管理系统的功能设计。
2. 掌握数据库设计方法。
3. 掌握分析类。
4. 掌握Java GUI设计方法。
5. 掌握Java中监听用户操作和数据库的处理。

■ 能力目标

1. 熟练使用软件工程的思想进行系统分析与设计。
2. 熟练使用Java编码并进行调试与测试。

■ 素质目标

1. 能够阅读科技文档和撰写分析文档。
2. 能够查阅JDK API。
3. 增强团队协作能力。

9.1 项目背景描述

随着我国市场经济的快速发展，各行业都呈现出生机勃勃的发展景象，其中餐饮业的发展尤为突出。随着餐饮企业规模和数量的不断增长，手工管理模式无论是在工作效率、人员成本还是提供决策信息方面都已难以适应现代化经营管理的要求，因此制约了整个餐饮业的规模化发展和整体服务水平的提升。随着社会各领域信息化建设的不断普及，餐饮业也开始不断注入信息化元素，将餐饮业务融入计算机管理，既节省人力资源，也提高了管理效率和工作效率，将餐饮业提升到一个新的阶段。

根据餐饮系统的流程，完成从用户登录到开台点菜，到结账收银，到统计一条线的信息化管理，使用计算机对餐饮企业信息进行管理，具有手工管理所无法比拟的优点。例如：检索迅速、查找方便、可靠性高、存储量大、保密性好、寿命长、成本低等。因此整个餐饮管理信息系统的研发内容就是开发一整套餐饮管理信息系统，实现餐饮业务的信息化。

9.2 系统需求分析

根据餐饮行业的特点和该企业的实际情况，本餐饮管理系统以餐饮业务为基础，突出管理，从专业角度出发，提供科学有效的管理模式。

能够针对中餐为其菜品的多样化和特色化的服务提供标准化的管理。

能够提供符合餐饮企业自身要求的较科学的标准化、流程化管理，解决餐饮行业专业人才欠缺的问题。

能够依据订餐、点菜、结账等环节的繁重化及复杂化问题，实现强化管理、降低成本、堵漏节流等效益。

能够针对企业的经营现状做出科学的分析，使得企业对市场的应变能力得到提高。

本餐饮管理系统采用 Java 语言进行开发，JDK 采用 1.7，开发工具使用 Eclipse，数据库使用 mySQL 5.0。

9.3 系统总体设计

根据该企业的具体情况，系统主要功能设计有六大部分，分别为员工管理、客户管理、餐台管理、菜品管理、点菜管理、结账管理，如图 9-1 所示。

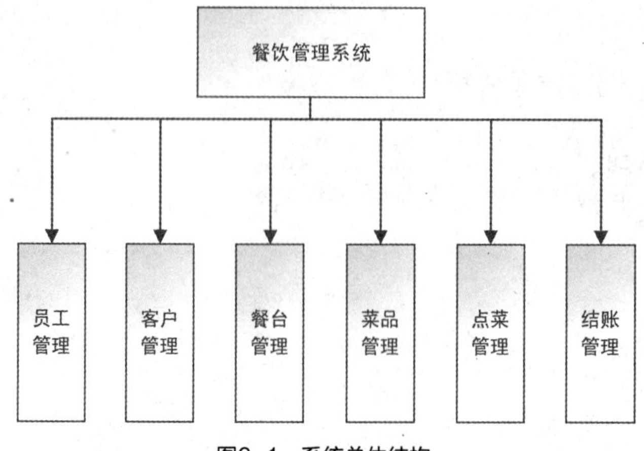

图9-1　系统总体结构

（1）员工管理：对员工实现增删改查。

（2）客户管理：对客户实现增删改查。

（3）餐台管理：对餐台实现增删改查。

（4）菜品管理：对菜品分类、菜品实现增删改查。

（5）点菜管理：服务员对某客户、某一空闲餐台实行开台，同时实现点菜，将餐台号与所点的菜品对应起来，分别显示出来，并记录开台时间。

（6）结账管理：收银员对某一餐台通过统计显示消费的菜品清单统计出消费金额，通过手动输入实收金额进行找零的计算，并显示，完成结账的操作，并记录成统计数据。

9.4 系统数据库设计

管理员信息表 user

列名	数据类型	长度	是否允许为空	是否为主键	说明
id	Int	10	no	yes	序号
username	varchar	20	no	no	用户名
password	varchar	20	no	no	密码

员工信息表

列名	数据类型	长度	是否允许为空	是否为主键	说明
id	Int	10	no	yes	序号
name	varchar	20	No	no	用户名
sex	varchar	2	yes	no	性别
birthday	datatime	8	yes	no	出生日期
identityID	varchar	18	yes	no	身份证号
address	varchar	40	yes	no	家庭住址
tel	varchar	11	yes	no	电话
position	varchar	4	no	no	职位
freeze	varchar	4	no	no	是否在职

客户信息表 customer

列名	数据类型	长度	是否允许为空	是否为主键	说明
id	Int	10	no	yes	序号
name	varchar	20	No	no	用户名
sex	varchar	4	yes	no	性别
company	varchar	20	yes	no	单位
tel	varchar	11	yes	no	电话
cardID	varchar	10	no	no	贵宾卡号

菜品分类表 category

列名	数据类型	长度	是否允许为空	是否为主键	说明
id	Int	10	no	yes	序号
name	varchar	20	no	no	名称
describ	varchar	20	yes	no	描述

菜品信息表

列名	数据类型	长度	是否允许为空	是否为主键	说明
id	Int	10	no	yes	序号
name	varchar	20	no	no	菜品名
categoryId	Int	10	no	no	类别编号
pic	Blob	50	no	no	图片
code	varchar	8	no	no	菜品代码
unit	varchar	4	yes	no	单位
price	datatime	6	yes	no	价格
status	varchar	4	yes	no	状态

餐台信息表 desk

列名	数据类型	长度	是否允许为空	是否为主键	说明
id	Int	10	no	yes	序号
no	varchar	8	no	yes	餐台编号
seating	Int	4	no	no	座位数
status	varchar	10	no	no	状态：已预订、就餐中、已结账

订单信息表 order

列名	数据类型	长度	是否允许为空	是否为主键	说明
id	Int	10	no	yes	序号
orderNo	varchar	20	no	yes	订单编号（当前日期时间 +4 位随机数）
deskId	Int	10	no	no	餐台号，外键
createtime	Date	40	no	no	就餐日期时间
money	double	6	no	no	金额
customerId	Int	10	no	no	客户编号
status	varchar	4	no	no	状态：已支付、未支付
number	Int	4	no	no	就餐人数

订单明细表 orderitem

列名	数据类型	长度	是否允许为空	是否为主键	说明
id	Int	10	no	yes	序号
orderId	Int	10	no	no	订单编号，外键
dishId	Int	10	no	no	菜品编号，外键
amount	double	4	no	no	菜品数量

9.5 系统界面分析与设计

1. 登录界面

如图 9-2 所示。

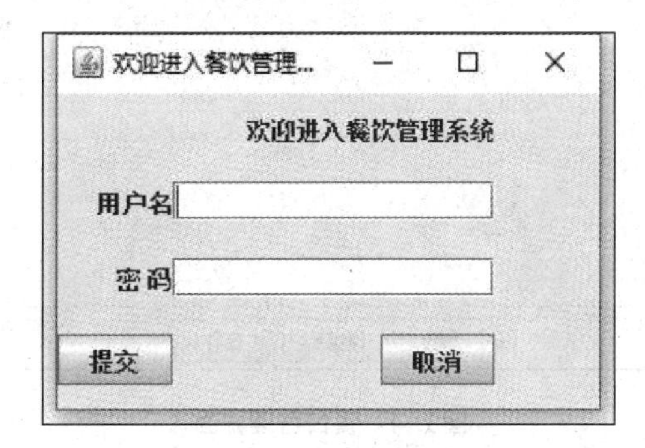

图9-2 登录界面

2. 系统主界面

如图 9-3 所示。

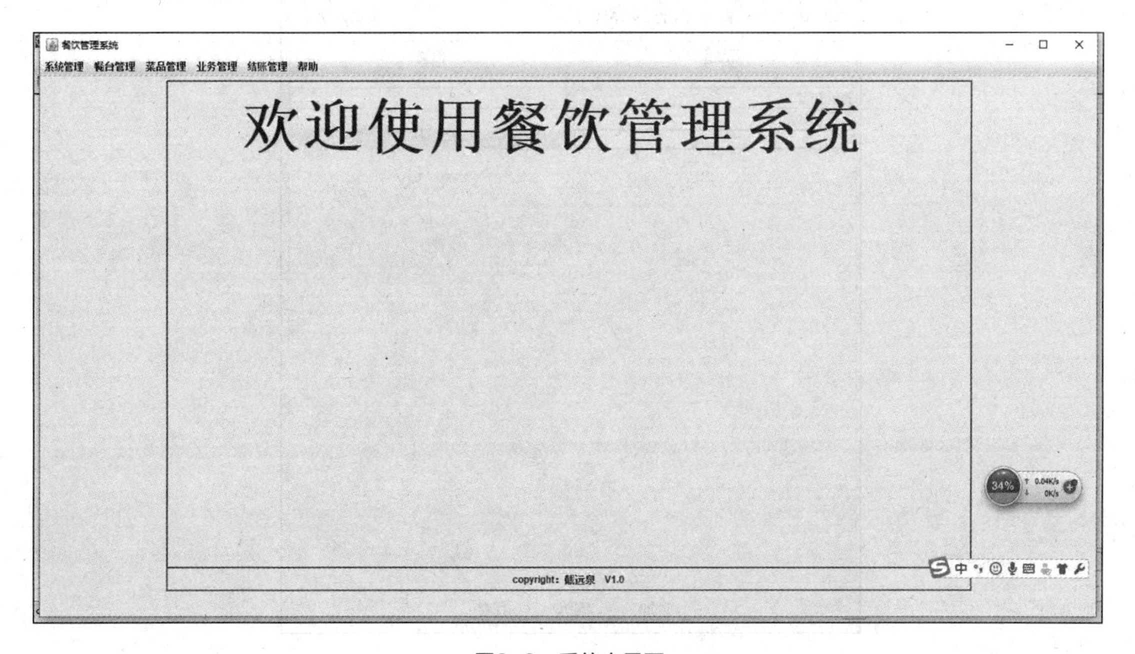

图9-3 系统主界面

3. 餐台管理界面

如图 9-4 所示。

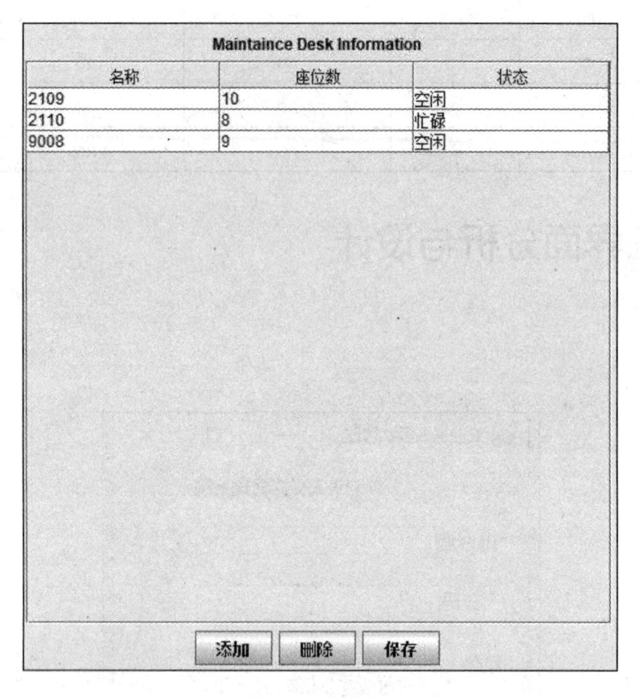

图9-4　餐台管理界面

4. 点菜管理界面

如图 9-5 所示。

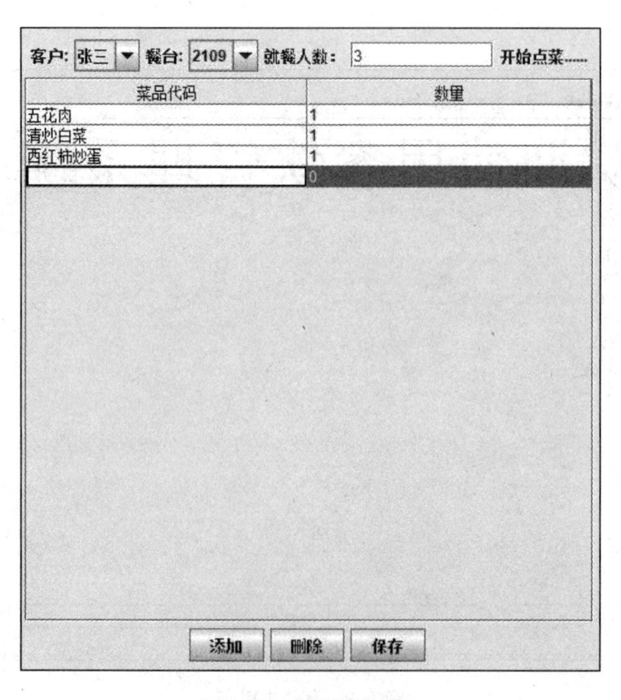

图9-5　点菜管理界面

5. 菜品管理界面

如图 9-6 所示。

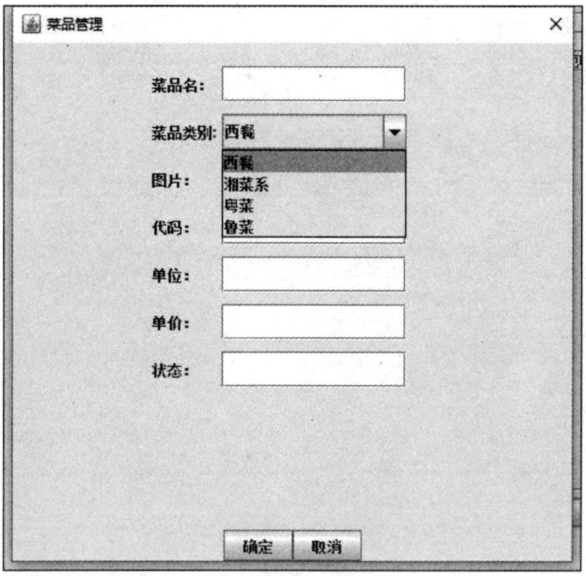

图9-6　菜品管理界面

6. 菜品分类管理界面

如图 9-7 所示。

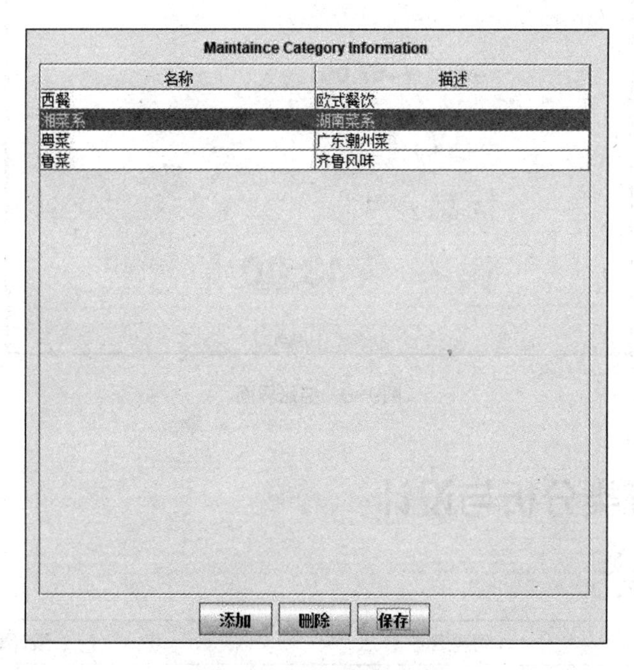

图9-7　菜品分类管理界面

7. 结账界面

如图 9-8、图 9-9 所示。

Maintaince Order Information

订单编号	餐台编号	创建日期	总金额	客户编号	状态	就餐人数
20180506	2110	20180506	457.0	李	未支付	10
18051270	2110	1805127	0.0	李	未支付	5
2018051...	2109	2018051...	0.0	张三	未支付	1
2018051...	2109	2018051...	0.0	张三	未支付	6
2018051...	2109	2018051...	0.0	张三	未支付	5

图9-8　查询未支付界面

结账管理

餐桌号：2110　客户:　李

应收:¥457.00

实收：500.00

优惠：0.00

找零：¥43.00

确定　取消

图9-9　结账界面

9.6　系统类分析与设计

9.6.1　实体类

类名	功能描述	设计要点
User.java	定义管理员信息	和管理员信息表中的信息一一对应
Employee.java	定义员工信息	和员工信息表中的信息一一对应
Customer.java	定义客户信息	和客户信息表中的信息一一对应

续表

类名	功能描述	设计要点
Desk.java	定义餐台信息	和餐台信息表中的信息一一对应
Category.java	定义菜品分类信息	和菜品分类信息表的信息一一对应
Dish.java	定义菜品信息	和菜品信息表中的信息一一对应
Order.java	定义开台信息	和订单信息表中的信息一一对应
OrderItem.java	定义点菜信息	和点菜信息表中的信息一一对应

9.6.2 边界类

类名	功能描述	设计要点
LoginFrame.java	用户登录界面	将用户登录名和密码与管理员信息表中的内容对比，正确则进入系统主界面，否则提示错误信息
MainFrame.java	系统主界面	提供系统功能菜单，并通过为各子菜单增加事件监听器以调用相应的功能模块
MainCatering.java	系统主程序	生成系统主程序
EmployeeManagePane2.java	员工管理界面	提供员工列表，并提供增删改查操作入口按钮
CustomerManagePane2.java	客户管理界面	提供客户列表，并提供增删改查操作入口按钮
DeskManagePane2.java	餐台管理界面	提供餐台列表，并提供增删改查操作入口按钮
CategoryManagePane2.java	菜品分类管理界面	提供菜品分类列表，并提供增删改查操作入口按钮
DishManagePane2.java	菜品管理界面	提供菜品列表，并提供增删改查操作入口按钮
DishesAddDialog.java	增加菜品界面	添加菜品对话框，保存记录时要检查数据的有效性，编号唯一、数据准确
OrderesManagePane2.java	开台管理界面	提供空餐台列表，选择客户、餐台，并生成订单
DishesOrderesManagePane2.java	点菜管理界面	提供全部菜品供客户挑选
ShowDishesDialog.java	显示该餐台所点菜品界面	显示菜品清单及总金额并提供修改增加和删除
DishesOrderesManagePane2.java	结账界面	显示菜品清单及总金额
GiveChangeDialog.java	找零界面	显示总金额、预付及找零

9.6.3 控制类

类名	功能描述	设计要点
JDBCConnection.java	数据库操作	主要用于数据库的连接、关闭
IBaseDAO.java	定义泛型接口	用于对实体类进行增删改查操作
UserDAOImpl.java	定义对管理员进行操作	继承 IBaseDAO.java，对管理员信息表进行 CRUD 操作
EmployeeDAOImpl.java	定义对员工进行操作	继承 IBaseDAO.java，对员工信息表进行 CRUD 操作
CustomerDAOImpl.java	定义对就餐区域进行操作	继承 IBaseDAO.java，对就餐区域信息表进行 CRUD 操作
DeskDAOImpl.java	定义对餐台进行操作	继承 IBaseDAO.java，对餐台信息表进行 CRUD 操作
CategoryDAOImpl.java	定义对菜品分类进行操作	
DishDAOImpl.java	定义对菜品进行操作	继承 IBaseDAO.java，对菜品信息表进行 CRUD 操作

续表

类名	功能描述	设计要点
OrderDAOImpl.java	定义对预订餐台进行操作	继承 IBaseDAO.java，对开台信息表进行 CRUD 操作
OrderItemDAOImpl.java	定义对点菜进行操作	继承 IBaseDAO.java，对点菜信息表进行 CRUD 操作

9.6.4 其他类

如图 9-10 所示。

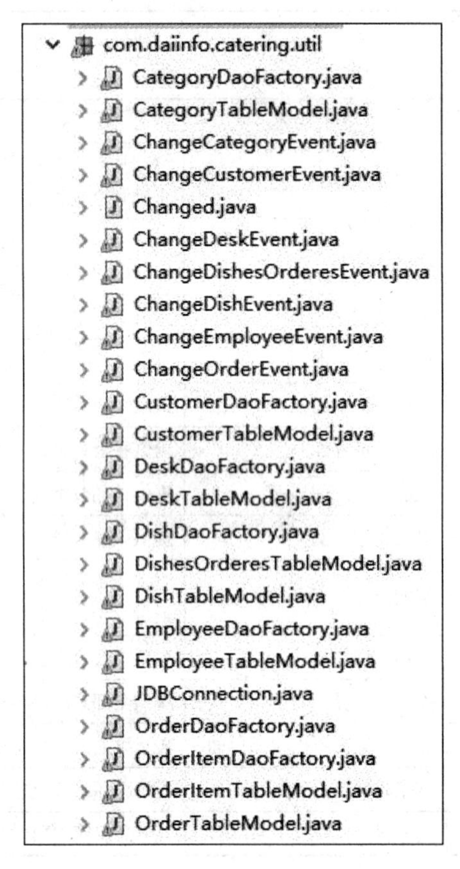

图9-10　功能类

9.7 系统功能的实现

9.7.1 系统登录窗口

```
package com.daiinfo.catering.frame;
import java.awt.BorderLayout;
import java.awt.Component;
import java.awt.Dimension;
import java.awt.GridBagConstraints;
import java.awt.GridBagLayout;
import java.awt.Toolkit;
import javax.swing.*;
```

```
import javax.swing.JFrame;
import javax.swing.JPanel;
class LoginFrame extends JPanel{
    private static final long serialVersionUID = 1834511718758119719L;
    static final int WIDTH = 300;
    static final int HEIGHT = 200;
    JFrame loginframe;
/*
 *按照网格组布局方式排列组件
  x指控件位于第几列;
  y指控件位于第几行;
  w指控件需要占几列;
  h指控件需要占几行。
 */
    public void add(Component c, GridBagConstraints constraints, int x, int y, int w, int h){
        constraints.gridx = x;
        constraints.gridy = y;
        constraints.gridwidth = w;
        constraints.gridheight = h;
        add(c, constraints);
    }
    LoginFrame(){
        loginframe = new JFrame("欢迎进入餐饮管理系统");
        loginframe.setDefaultCloseOperation(JFrame.EXIT_ON_CLOSE);
        GridBagLayout lay = new GridBagLayout();
        setLayout(lay); // 这里panel的布局使用grid
        loginframe.add(this, BorderLayout.WEST);// frame的布局使用BorderLayout
        loginframe.setSize(WIDTH, HEIGHT);
        Toolkit kit = Toolkit.getDefaultToolkit();
        Dimension screenSize = kit.getScreenSize(); // 系统对象获取工具
        int width = screenSize.width;
        int height = screenSize.height;
        int x = (width - WIDTH) / 2;
        int y = (height - HEIGHT) / 2;
        loginframe.setLocation(x, y);
        JButton ok = new JButton("提交");
        JButton cancel = new JButton("取消");
        JLabel title = new JLabel("欢迎进入餐饮管理系统");
        JLabel name = new JLabel("用户名");
        JLabel password = new JLabel("密 码");
        final JTextField nameinput = new JTextField(15);
        final JTextField passwordinput = new JTextField(15);
        GridBagConstraints constraints = new GridBagConstraints();
        constraints.fill = GridBagConstraints.NONE;
        constraints.anchor = GridBagConstraints.EAST;
        constraints.weightx = 3;
        constraints.weighty = 4;
        add(title, constraints, 0, 0, 3, 1); // 使用网格组布局添加控件
        add(name, constraints, 0, 1, 1, 1);
        add(password, constraints, 0, 2, 1, 1);
        add(nameinput, constraints, 2, 1, 1, 1);
```

```
        add(passwordinput, constraints, 2, 2, 1, 1);
        add(ok, constraints, 0, 3, 1, 1);
        add(cancel, constraints, 2, 3, 1, 1);
        loginframe.setVisible(true);
    }
    public static void main(String[] args) {
        LoginFrame login = new LoginFrame();
    }
}
```

9.7.2 系统主窗口

```
package com.daiinfo.catering.frame;
import java.awt.FlowLayout;
import java.awt.event.ActionEvent;
import java.awt.event.ActionListener;
import javax.swing.JFrame;
import javax.swing.JMenu;
import javax.swing.JMenuBar;
import javax.swing.JMenuItem;
import javax.swing.JPanel;
import javax.swing.JTextField;
import com.daiinfo.catering.pane.*;
public class MainFrame extends JFrame  {
    JPanel panel;
    JMenuBar bar;
    JMenu systemMenu,deskMenu,dishesMenu,orderMenu,checkoutMenu,helpMenu;
    JMenuItem emplManage,custManage, pwdManage,categoryManage,dishesManage,
    aboutmeManage,orderManage,checkoutManage,orderItemManage,deskManage;
    FlowLayout layout = new FlowLayout();

    EmployeeManagePane2 emplManaPane2;
    //CustomerManagePane custManaPane;
    CustomerManagePane2 custManaPane2;
    PasswordManagePane pwdManaPane;
    DeskManagePane deskManaPane;
    DeskManagePane2 deskManaPane2;

    CategoryManagePane2 cateManaPane2;
    DishesManagePane dishManaPane;
    DishManagePane2 dishManaPane2;

    OrderManagePane orderManaPane;
    OrderesManagePane2 orderesManaPane2;
    DishesOrderesManagePane2 DishesOrderesManaPane2;
    DeskManagePane3 deskManaPane3;
    OrderItemManagePane orderItemManaPane;
    CheckoutManagePane checkoutManaPane;
    AboutmeManagePane aboutmeManaPane;
    MainFrame() {
        init();
        //setBounds(100, 100, 200, 200);
```

```
        setSize(1366,728);
        setVisible(true);
        setDefaultCloseOperation(JFrame.EXIT_ON_CLOSE);
    }
    public void init() {
        bar = new JMenuBar();
        systemMenu = new JMenu("系统管理");
        deskMenu=new JMenu("餐台管理");
        dishesMenu=new JMenu("菜品管理");
        orderMenu = new JMenu("业务管理");
        checkoutMenu= new JMenu("结账管理");
        helpMenu=new JMenu("帮助");


        emplManage = new JMenuItem("员工管理");
        custManage= new JMenuItem("客户管理");
        pwdManage = new JMenuItem("修改密码");

        deskManage=new JMenuItem("餐台管理");

        categoryManage=new JMenuItem("分类管理");
        dishesManage=new JMenuItem("菜品管理");

        orderManage=new JMenuItem("开台管理");
        orderItemManage=new JMenuItem("点菜管理");

        checkoutManage=new JMenuItem("结账管理");

        aboutmeManage=new JMenuItem("关于我们");

        bar.add(systemMenu);
        bar.add(deskMenu);
        bar.add(dishesMenu);
        bar.add(orderMenu);
        bar.add(checkoutMenu);
        bar.add(helpMenu);

        //系统管理菜单项
        systemMenu.add(emplManage);
        systemMenu.add(custManage);
        systemMenu.add(pwdManage);

        //餐台管理菜单项
        deskMenu.add(deskManage);

        //菜品管理菜单项
        dishesMenu.add(categoryManage);
        dishesMenu.add(dishesManage);

        //业务管理菜单项
        orderMenu.add(orderManage);
```

```
orderMenu.add(orderItemManage);

//结账管理菜单项
checkoutMenu.add(checkoutManage);

//帮助菜单项
helpMenu.add(aboutmeManage);

panel = new JPanel();

emplManaPane2=new EmployeeManagePane2();
//custManaPane=new CustomerManagePane();
custManaPane2=new CustomerManagePane2();
pwdManaPane=new PasswordManagePane();

deskManaPane=new DeskManagePane();
deskManaPane2=new DeskManagePane2();
deskManaPane3=new DeskManagePane3();

cateManaPane2=new CategoryManagePane2();
dishManaPane=new DishesManagePane();
dishManaPane2=new DishManagePane2();
orderesManaPane2=new OrderesManagePane2();
DishesOrderesManaPane2=new DishesOrderesManagePane2();
orderItemManaPane=new OrderItemManagePane();
checkoutManaPane=new CheckoutManagePane();
aboutmeManaPane=new AboutmeManagePane();

emplManage.addActionListener(new ActionListener(){
    @Override
    public void actionPerformed(ActionEvent e) {
        // TODO Auto-generated method stub
        panel.removeAll();
        panel.add("ty", emplManaPane2);// 切换代码
        panel.validate();
        repaint();
    }

});

custManage.addActionListener(new ActionListener(){
    @Override
    public void actionPerformed(ActionEvent e) {
        // TODO Auto-generated method stub
        panel.removeAll();
        panel.add("客户管理", custManaPane2);// 切换代码
        panel.validate();
        repaint();
    }

});
```

```
pwdManage.addActionListener(new ActionListener(){
    @Override
    public void actionPerformed(ActionEvent e) {
        // TODO Auto-generated method stub
        panel.removeAll();
        panel.add("pmg",pwdManaPane);
        panel.validate();
        repaint();
    }

});

deskManage.addActionListener(new ActionListener(){
    @Override
    public void actionPerformed(ActionEvent e) {
        // TODO Auto-generated method stub
        panel.removeAll();
        panel.add("dmp",deskManaPane2);
        panel.validate();
        repaint();
    }

});
categoryManage.addActionListener(new ActionListener(){
    @Override
    public void actionPerformed(ActionEvent e) {
        // TODO Auto-generated method stub
        panel.removeAll();
        panel.add("cmp", cateManaPane2);
        panel.validate();
        repaint();
    }

});

dishesManage.addActionListener(new ActionListener(){
    @Override
    public void actionPerformed(ActionEvent arg0) {
        // TODO Auto-generated method stub
        panel.removeAll();
        panel.add("jk", dishManaPane2);
        panel.validate();
        repaint();
    }
});

orderManage.addActionListener(new ActionListener(){
    @Override
    public void actionPerformed(ActionEvent arg0) {
        // TODO Auto-generated method stub
```

```
                panel.removeAll();
                panel.add("omp", deskManaPane3);
                panel.validate();
                repaint();
            }
        });

        orderItemManage.addActionListener(new ActionListener(){
            @Override
            public void actionPerformed(ActionEvent arg0) {
                // TODO Auto-generated method stub
                panel.removeAll();
                panel.add("oimp", DishesOrderesManaPane2);
                panel.validate();
                repaint();
            }
        });

        checkoutManage.addActionListener(new ActionListener(){
            @Override
            public void actionPerformed(ActionEvent arg0) {
                // TODO Auto-generated method stub
                panel.removeAll();
                panel.add("jk", orderesManaPane2);
                panel.validate();
                repaint();
            }
        });

        aboutmeManage.addActionListener(new ActionListener(){
            @Override
            public void actionPerformed(ActionEvent e) {
                // TODO Auto-generated method stub
                panel.removeAll();
                panel.add("gh", aboutmeManaPane);// 切换代码
                panel.validate();
                repaint();
            }

        });

        panel.setLayout(layout);
        add(panel);
        setJMenuBar(bar);
    }
}
```

9.7.3 系统主程序

```
package com.daiinfo.catering.frame;
public class MainCatering {
    public static void main(String[] args) {
```

```
        // TODO Auto-generated method stub
        MainFrame mf = new MainFrame();
        mf.setTitle("餐饮管理系统");
    }
}
```

9.7.4 菜品分类管理

1. 运行效果

如图 9-11 所示。

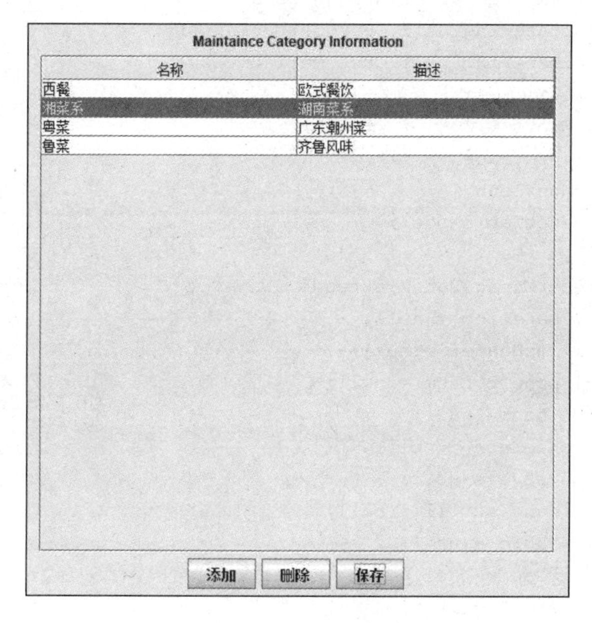

图9-11 运行效果

2. 界面代码

CategoryManagePane2.java

```
package com.daiinfo.catering.pane;
import java.awt.BorderLayout;
import java.awt.Color;
import java.awt.Container;
import java.awt.Dimension;
import java.awt.FlowLayout;
import java.awt.event.ActionEvent;
import java.awt.event.ActionListener;
import java.util.ArrayList;
import java.util.List;
import javax.swing.BorderFactory;
import javax.swing.JButton;
import javax.swing.JFrame;
import javax.swing.JLabel;
import javax.swing.JOptionPane;
import javax.swing.JPanel;
```

```
import javax.swing.JScrollPane;
import javax.swing.JTable;
import javax.swing.table.DefaultTableColumnModel;
import com.daiinfo.category.dialog.CustomerAddDialog;
import com.daiinfo.category.dialog.EmployeeAddDialog;
import com.daiinfo.catering.dao.IBaseDao;
import com.daiinfo.catering.entity.Category;
import com.daiinfo.catering.entity.Customer;
import com.daiinfo.catering.util.CategoryDaoFactory;
import com.daiinfo.catering.util.CategoryTableModel;
import com.daiinfo.catering.util.ChangeCategoryEvent;
import com.daiinfo.catering.util.ChangeCustomerEvent;
import com.daiinfo.catering.util.CustomerDaoFactory;
import com.daiinfo.catering.util.CustomerTableModel;
/*import product.ChangeEvent;
import product.DaoFactory;
import product.DaoInterface;
import product.ProductBean;
import product.ProductTableModel;
*/
public class CategoryManagePane2 extends JPanel {
    private JPanel panelTop = null;
    private JLabel labHeader = null;
    private JPanel panelBottom = null;
    private JButton add = null;
    private JButton delete = null;
    private JButton save = null;
    private JScrollPane scroll = null;
    private JTable table = null;
    private CategoryTableModel model = null;
    private List listCategory = null;
    public CategoryTableModel getModel() {
        if (null == model) {
            model = new CategoryTableModel(listCategory);
            // 给model添加一个监听，当修改的时候将触发该事件，代表事件的类是ChangeEvent
            model.addTableModelListener(new ChangeCategoryEvent(model));
            return model;
        }
        return model;
    }
    public JLabel getLabHeader() {
        if (null == labHeader) {
            labHeader = new JLabel("Maintaince Category Information");
            return labHeader;
        }
        return labHeader;
    }
    public JTable getTable() {
        if (null == table) {
            table = new JTable(getModel());
            table.setEnabled(true);
```

```
        table.setRowSelectionAllowed(true);
        //table.setBackground(Color.blue);
        table.setSelectionForeground(Color.yellow);
        table.setSelectionBackground(Color.GRAY);
        //table.setForeground(Color.red);
        /**
         * 隐藏第一列ID，不显示出来
         */
        DefaultTableColumnModel dcm = (DefaultTableColumnModel) table.getColumnModel();
        dcm.getColumn(0).setMinWidth(0);
        dcm.getColumn(0).setMaxWidth(0);
        return table;
    }
    return table;
}
public JScrollPane getPanTable() {
    if (null == scroll) {
        scroll = new JScrollPane();
        scroll.setViewportView(getTable());
        return scroll;
    }
    return scroll;
}
public JPanel getPanelTop() {
    if (null == panelTop) {
        panelTop = new JPanel();
        panelTop.setLayout(new FlowLayout(FlowLayout.CENTER));
        panelTop.add(getLabHeader());
        return panelTop;
    }
    return panelTop;
}
public JPanel getPanelBottom() {
    if (null == panelBottom) {
        panelBottom = new JPanel();
        panelBottom.setLayout(new FlowLayout(FlowLayout.CENTER));
        panelBottom.add(getAdd());
        panelBottom.add(getDelete());
        panelBottom.add(getSave());
        return panelBottom;
    }
    return panelBottom;
}
public JButton getAdd() {
    /**
     * 点该按钮的时候调用addProduct()方法，在数据源（listProduct）将增加一个元素，没设值前都是null
     */
    if (null == add) {
        add = new JButton("添加");
        add.addActionListener(new ActionListener() {
            public void actionPerformed(ActionEvent e) {
```

```
                addCategory();
            }
        });
        return add;
    }
    return add;
}
public JButton getDelete() {
    if (null == delete) {
        delete = new JButton("删除");
        delete.addActionListener(new ActionListener() {
            public void actionPerformed(ActionEvent e) {
                /**
                 * 支持一次选中多行后删除
                 */
                int[] rows = getTable().getSelectedRows();
                if (rows.length > 0) {
                    int flag = JOptionPane.showConfirmDialog(null, "确定删除？");
                    if (flag == JOptionPane.YES_OPTION)
                        deleteCategory();
                } else
                    JOptionPane.showMessageDialog(null, "请选择要删除的行！");
            }
        });
        return delete;
    }
    return delete;
}
public JButton getSave() {
    if (null == save) {
        save = new JButton("保存");
        save.addActionListener(new ActionListener() {
            public void actionPerformed(ActionEvent e) {
                saveCategory();
                JOptionPane.showMessageDialog(null, "更新成功！");
            }
        });
        return save;
    }
    return save;
}
public void addCategory() {
    Category cust = new Category();
    getModel().addRow(getTable().getSelectedRow(), cust);
}
public void saveCategory() {
    IBaseDao dao = CategoryDaoFactory.getDao();
    List changeList = getModel().getChangeList();
    // 如果有修改过就调用update方法
    if (changeList.size() > 0) {
        dao.update(changeList);
```

```
            changeList.clear();
        }
        List newRow = getModel().getNewRow();
        // 如果是新增就调用saveList，支持一次增加多行
        if (newRow.size() > 0) {
            dao.saveList(newRow);
            getModel().setList(dao.getList());
            getTable().updateUI();
            newRow.clear();
        }
    }
    public void deleteCategory() {
        /**
         * 支持一次删除多行，先获得所有选中的行，然后按照行数取得Product实例，
         * 放进一个list，然后传给操作数据库的deleteList方法
         */
        int[] rows = getTable().getSelectedRows();
        ArrayList list = new ArrayList();
        IBaseDao dao = CategoryDaoFactory.getDao();
        for (int i = rows.length - 1; i >= 0; i--) {
            list.add(getModel().getRow(rows[i]));
            getModel().deleteRow(rows[i]);
        }
        dao.deleteList(list);
        getTable().updateUI();
        list.clear();
    }
    public void initData() {
        /**
         * 初始化数据源，从数据库里把数据拿出来，然后它会调用 getValueAt方法一个单元格一个单元格
来设值，让它显示出来
         */
        listCategory = new ArrayList();
        IBaseDao dao = CategoryDaoFactory.getDao();
        listCategory = dao.getList();
    }
    public CategoryManagePane2() {
        initData();
        this.setLayout(new BorderLayout());
        add(getPanelTop(), BorderLayout.NORTH);
        add(getPanelBottom(), BorderLayout.SOUTH);
        add(getPanTable(), BorderLayout.CENTER);
        this.setSize(new Dimension(1024,800));
    }
}
```

JDBConnection.java

```
package com.daiinfo.catering.util;
import java.sql.Connection;
import java.sql.DriverManager;
import java.sql.ResultSet;
```

```
import java.sql.SQLException;
import com.mysql.jdbc.Statement;
public class JDBConnection {
    /**
     *
     */
    private static Connection conn;
    private static final String DRIVER = "com.mysql.jdbc.Driver";
    private static final String URL = "jdbc:mysql://localhost:3306/catering";
    private static final String USRENAME = "root";
    private static final String PASSWORD = "123456";
    static {
        try {
            Class.forName(DRIVER);
        } catch (ClassNotFoundException e) {
            e.printStackTrace();
        }
    }
    public static void main(String[] args) {
        System.out.print(getConn());
    }
    public static Connection getConn() {
        Connection conn = null;
        try {
            conn = DriverManager.getConnection(URL, USRENAME, PASSWORD);
        } catch (SQLException e) {
            e.printStackTrace();
        }
        return conn;
    }
    public static void close(Connection conn, Statement st, ResultSet rs) {
        try {
            if (rs != null) {
                rs.close();
            }
            if (st != null) {
                st.close();
            }
            if (conn != null) {
                conn.close();
            }
        } catch (Exception e) {
            // TODO Auto-generated catch block
            e.printStackTrace();
        }
    }
}
```

Category.java

```
package com.daiinfo.catering.entity;
import java.io.Serializable;
```

```
public class Category  implements Serializable{
    int id;
    String name;
    String describ;
    public int getId() {
        return id;
    }
    public void setId(int id) {
        this.id = id;
    }
    public String getName() {
        return name;
    }
    public void setName(String name) {
        this.name = name;
    }
    public String getDescrib() {
        return describ;
    }
    public void setDescrib(String describ) {
        this.describ = describ;
    }
}
```

IBaseDao.java

```
package com.daiinfo.catering.dao;
import java.util.ArrayList;
import java.util.List;
public interface IBaseDao {
    /**
     *
     */
    public List getList();
    public void saveList(List list);
    public void deleteList(List list);
    public void update(List list);
}
```

CategoryDaoImpl.java

```
package com.daiinfo.catering.dao;
import java.sql.Connection;
import java.sql.PreparedStatement;
import java.sql.ResultSet;
import java.sql.SQLException;
import java.util.ArrayList;
import java.util.Iterator;
import java.util.List;
import javax.swing.JOptionPane;
import com.daiinfo.catering.entity.Category;
import com.daiinfo.catering.entity.Customer;
```

```
import com.daiinfo.catering.util.Changed;
import com.daiinfo.catering.util.JDBConnection;
public class CategoryDaoImpl implements IBaseDao {
    Category cate;
    public Category getCategoryById(int id) {
        cate = new Category();
        Connection conn = JDBConnection.getConn();
        String s1 = "select * from category where id=?";
        PreparedStatement ps = null;
        ResultSet rs = null;
        /* List list = new ArrayList(); */
        try {
            ps = conn.prepareStatement(s1);
            ps.setInt(1, id);
            rs = ps.executeQuery();
        } catch (SQLException e) {
            JOptionPane.showMessageDialog(null, "①取出category全部数据出错！");
            e.printStackTrace();
        }
        try {
            rs.next();
            cate.setId(rs.getInt("id"));
            cate.setName(rs.getString("name"));
            cate.setDescrib(rs.getString("describ"));
        } catch (SQLException e) {
            JOptionPane.showMessageDialog(null, "②取出category数据出错！");
            e.printStackTrace();
        } finally {
            try {
                rs.close();
                ps.close();
                conn.close();
            } catch (SQLException e) {
                JOptionPane.showMessageDialog(null, "关闭数据连接时出错！");
                e.printStackTrace();
            }
        }
        return cate;
    }

    public Category getCategoryByName(String name){
        cate = new Category();
        Connection conn = JDBConnection.getConn();
        String s1 = "select * from category where name=?";
        PreparedStatement ps = null;
        ResultSet rs = null;
        /* List list = new ArrayList(); */
        try {
            ps = conn.prepareStatement(s1);
            ps.setString(1, name);
            rs = ps.executeQuery();
```

```
        } catch (SQLException e) {
            JOptionPane.showMessageDialog(null, "①取出category全部数据出错！");
            e.printStackTrace();
        }
        try {
            rs.next();
            cate.setId(rs.getInt("id"));
            cate.setName(rs.getString("name"));
            cate.setDescrib(rs.getString("describ"));
        } catch (SQLException e) {
            JOptionPane.showMessageDialog(null, "②取出category数据出错！");
            e.printStackTrace();
        } finally {
            try {
                rs.close();
                ps.close();
                conn.close();
            } catch (SQLException e) {
                JOptionPane.showMessageDialog(null, "关闭数据连接时出错！");
                e.printStackTrace();
            }
        }
        return cate;
    }
    @Override
    public List getList() {
        // TODO Auto-generated method stub
        Connection conn = JDBConnection.getConn();
        String s1 = "select * from category order by id asc";
        PreparedStatement ps = null;
        ResultSet rs = null;
        List list = new ArrayList();
        try {
            ps = conn.prepareStatement(s1);
            rs = ps.executeQuery();
        } catch (SQLException e) {
            System.out.println("取出全部数据出错！");
            JOptionPane.showMessageDialog(null, "取出全部数据出错！");
            e.printStackTrace();
        }
        try {
            while (rs.next()) {
                Category cate = new Category();
                cate.setId(rs.getInt(1));
                cate.setName(rs.getString(2));
                cate.setDescrib(rs.getString(3));
                list.add(cate);
            }
        } catch (SQLException e) {
            JOptionPane.showMessageDialog(null, "取出全部数据出错！");
            e.printStackTrace();
```

```
        } finally {
            try {
                rs.close();
                ps.close();
                conn.close();
            } catch (SQLException e) {
                JOptionPane.showMessageDialog(null, "关闭数据连接时出错！");
                e.printStackTrace();
            }
        }
        return list;
    }
    @Override
    public void saveList(List list) {
        // TODO Auto-generated method stub
        String s1 = "";
        Connection conn = JDBConnection.getConn();
        PreparedStatement ps = null;
        try {
            Iterator it = list.iterator();
            while (it.hasNext()) {
                Category cate = (Category) it.next();
                int id = cate.getId();
                String name = cate.getName();
                String describ = cate.getDescrib();
                s1 = "insert into category(name,describ) values(?,?)";
                ps = conn.prepareStatement(s1);
                ps.setString(1, name);
                ps.setString(2, describ);
                ps.executeUpdate();
            }
        } catch (SQLException e) {
            System.out.println("添加数据时出错！");
            JOptionPane.showMessageDialog(null, "添加数据时出错！");
            e.printStackTrace();
        } finally {
            try {
                ps.close();
                conn.close();
            } catch (SQLException e) {
                JOptionPane.showMessageDialog(null, "关闭数据连接时出错！");
                e.printStackTrace();
            }
        }
    }
    @Override
    public void deleteList(List list) {
        // TODO Auto-generated method stub
        String str = "delete from category where id=?";
        int id = 0;
        Iterator it = list.iterator();
```

```
        Connection conn = JDBConnection.getConn();
        PreparedStatement ps = null;
        try {
            while (it.hasNext()) {
                id = ((Category) it.next()).getId();
                ps = conn.prepareStatement(str);
                ps.setInt(1, id);
                ps.executeUpdate();
            }
        } catch (SQLException e) {
        } finally {
            try {
                ps.close();
                conn.close();
            } catch (SQLException e) {
                JOptionPane.showMessageDialog(null, "关闭数据连接时出错！");
                e.printStackTrace();
            }
        }
    }
    @Override
    public void update(List list) {
        // TODO Auto-generated method stub
        Connection conn = JDBConnection.getConn();
        PreparedStatement ps = null;
        Iterator it = list.iterator();
        int id = 0;
        int col = 0;
        String value = "";
        String str = "";
        try {
            while (it.hasNext()) {
                Changed ch = (Changed) it.next();
                id = ch.getId();
                col = ch.getCol();
                value = ch.getValue();
                switch (col) {
                case 1:
                    str = "update category set name=? where id=?";
                    break;
                case 2:
                    str = "update category set describ=? where id=?";
                    break;
                }
                ps = conn.prepareStatement(str);
                ps.setString(1, value);
                ps.setInt(2, id);
                ps.executeUpdate();
            }
        } catch (SQLException e) {
            JOptionPane.showMessageDialog(null, "修改数据时出错！");
```

```
            e.printStackTrace();
        } finally {
            try {
                ps.close();
                conn.close();
            } catch (SQLException e) {
                JOptionPane.showMessageDialog(null, "关闭数据连接时出错! ");
                e.printStackTrace();
            }
        }
    }
}
```

Changed.java

```
package com.daiinfo.catering.util;
public class Changed {
    /**

    */
    private int id;
    private int col;
    private String value;
    public int getId() {
        return id;
    }
    public void setId(int id) {
        this.id = id;
    }
    public String getValue() {
        return value;
    }
    public void setValue(String value) {
        this.value = value;
    }
    public int getCol() {
        return col;
    }
    public void setCol(int col) {
        this.col = col;
    }
}
```

ChangeCategoryEvent.java

```
package com.daiinfo.catering.util;
import java.util.ArrayList;
import java.util.List;
import javax.swing.event.TableModelEvent;
import javax.swing.event.TableModelListener;
public class ChangeCategoryEvent implements TableModelListener {
    /**

    */
```

```
        /**
         * 监听table被改动的事件，主要目的是用来记录被修改过的值，这样做可以一次任意行地修改值，修改一个单元格的值就记录一次
         * 主要记录id、新值、列数
         */
        CategoryTableModel model = null;
        public ChangeCategoryEvent(CategoryTableModel model) {
            this.model = model;
        }
        List list = model.getChangeList();
        int id = 0;
        String value = "";
        public void tableChanged(TableModelEvent arg0) {
            int row = arg0.getFirstRow();
            int col = arg0.getColumn();
            if (col != -1) {
                Changed cp = new Changed();
                id = ((Integer) model.getValueAt(row, 0)).intValue();
                if (id != 0) {
                    value = model.getValueAt(row, col).toString();
                    cp.setId(id);
                    cp.setCol(col);
                    cp.setValue(value);
                    list.add(cp);
                }
            }
        }
    }
```

CategoryTableModel.java

```
    package com.daiinfo.catering.util;
    import java.util.ArrayList;
    import java.util.Iterator;
    import java.util.List;
    import javax.swing.table.AbstractTableModel;
    import com.daiinfo.catering.entity.Category;
    import com.daiinfo.catering.entity.Customer;
    public class CategoryTableModel extends AbstractTableModel {
        /**
         *
         * changeList用来存放被修改过的数据值，这样做是为了一次修改多行多值，保存的对象是ChangedProduct，只记录被修改过的值
         */
        private static List changeList = new ArrayList();
        private List list = new ArrayList();
        private String[] column = { "编号", "名称", "描述" };
        public CategoryTableModel() {
        }
        public CategoryTableModel(List list) {
            this();
            setList(list);
        }
```

```
    public int getColumnCount() {
        return column.length;
    }
    public int getRowCount() {
        return list.size();
    }
    /**
     * getValueAt方法就是使得数据在Table显示出来，给每个单元格设值
     */
    public Object getValueAt(int arg0, int arg1) {
        Category cate = (Category) list.get(arg0);
        return getPropertyValueByCol(cate, arg1);
    }
    public void addRow(int index, Category cate) {
        if (index < 0 || index > list.size() - 1) {
            list.add(cate);
            fireTableRowsInserted(list.size(), list.size());
        } else {
            list.add(index + 1, cate);
            fireTableRowsInserted(index, index);
        }
    }
    public boolean deleteRow(int index) {
        if (index >= 0 && index < list.size()) {
            list.remove(index);
            fireTableRowsDeleted(index, index);
            return true;
        } else
            return false;
    }
    public boolean saveRow(int index,  Category cate) {
        if (index >= 0 && index < list.size()) {
            list.set(index, cate);
            fireTableRowsUpdated(index, index);
            return true;
        } else
            return false;
    }
    public Category getRow(int index) {
        if (index >= 0 && index < list.size()) {
            return (Category) list.get(index);
        } else
            return null;
    }
    public List getNewRow() {
        List list = new ArrayList();
        List listProduct = getList();
        Iterator it = listProduct.iterator();
        while (it.hasNext()) {
            Category cate = new Category();
            cate = (Category) it.next();
            if (cate.getId() == 0) {
                list.add(cate);
```

```
            }
        }
        return list;
    }
    public List getList() {
        return list;
    }
    public void setList(List list) {
        this.list = list;
        fireTableDataChanged();
    }
    public String getColumnName(int i) {
        return column[i];
    }
    public void setColumn(String[] column) {
        this.column = column;
    }
    public Object getPropertyValueByCol(Category cate, int col) {
        switch (col) {
        case 0:
            return cate.getId();
        case 1:
            return cate.getName();
        case 2:
            return cate.getDescrib();
        }
        return null;
    }
    public void setPropertyValueByCol(Category cate, String value, int col) {
        switch (col) {
        case 1:
            cate.setName(value);
            break;
        case 2:
            cate.setDescrib(value);
            break;
        }
        fireTableDataChanged();
    }
    public boolean isCellEditable(int row, int column) {
        return true;
    }
    /**
     * setValueAt方法是使增加或修改值的时候生效，aValue就是你在单元格填的值，要把这些值保存到数
据源中
     */
    public void setValueAt(Object aValue, int rowIndex, int columnIndex) {
        Category cate = (Category) list.get(rowIndex);
        setPropertyValueByCol(cate, aValue.toString(), columnIndex);
        this.fireTableCellUpdated(rowIndex, columnIndex);
    }
```

```
    public static List getChangeList() {
        return changeList;
    }
    public static void setChangeList(List changeList) {
        CategoryTableModel.changeList = changeList;
    }
}
```

CategoryDaoFactory.java

```
package com.daiinfo.catering.util;
import com.daiinfo.catering.dao.CategoryDaoImpl;
import com.daiinfo.catering.dao.CustomerDaoImpl;
import com.daiinfo.catering.dao.IBaseDao;
public class CategoryDaoFactory {
    /**
     *
     */
    synchronized public static IBaseDao getDao() {
        IBaseDao dao = null;
        if (dao == null) {
            dao = new CategoryDaoImpl();
            return dao;
        }
        return dao;
    }
}
```

9.7.5 菜品管理

1. 运行效果

如图 9-12 所示。

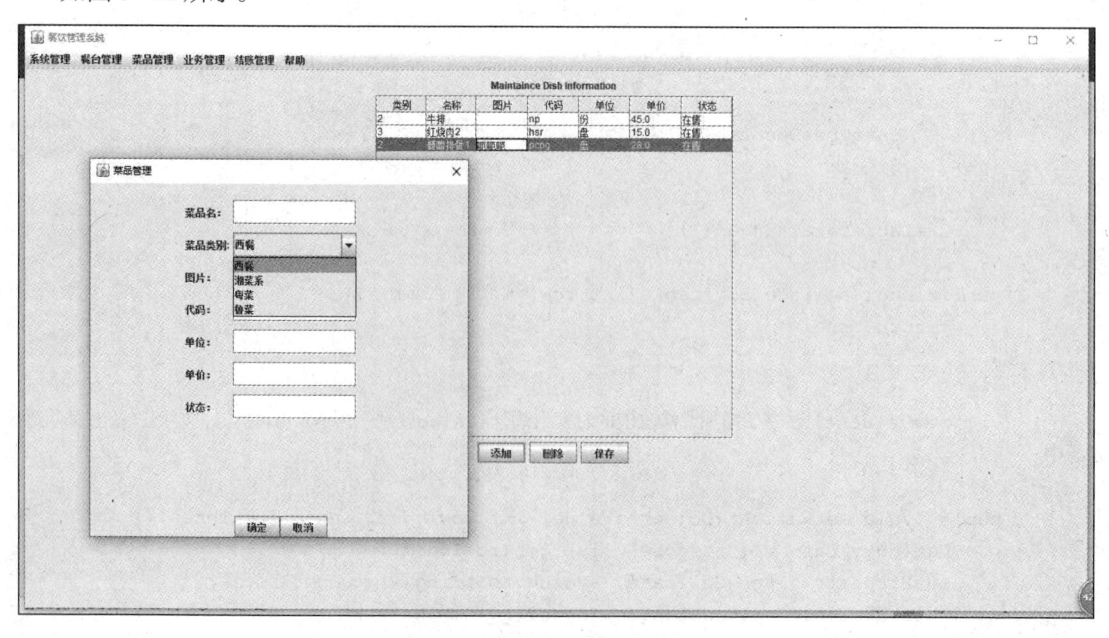

图9-12　菜品管理运行效果

2. 菜品管理面板类 DishManagePane2.java

```
package com.daiinfo.catering.pane;
import java.awt.BorderLayout;
import java.awt.Color;
import java.awt.Container;
import java.awt.Dimension;
import java.awt.FlowLayout;
import java.awt.event.ActionEvent;
import java.awt.event.ActionListener;
import java.util.ArrayList;
import java.util.List;
import javax.swing.BorderFactory;
import javax.swing.JButton;
import javax.swing.JFrame;
import javax.swing.JLabel;
import javax.swing.JOptionPane;
import javax.swing.JPanel;
import javax.swing.JScrollPane;
import javax.swing.JTable;
import javax.swing.table.DefaultTableColumnModel;
import com.daiinfo.category.dialog.CustomerAddDialog;
import com.daiinfo.category.dialog.DishesAddDialog;
import com.daiinfo.category.dialog.EmployeeAddDialog;
import com.daiinfo.catering.dao.IBaseDao;
import com.daiinfo.catering.entity.Category;
import com.daiinfo.catering.entity.Customer;
import com.daiinfo.catering.entity.Dish;
import com.daiinfo.catering.util.CategoryDaoFactory;
import com.daiinfo.catering.util.CategoryTableModel;
import com.daiinfo.catering.util.ChangeCategoryEvent;
import com.daiinfo.catering.util.ChangeCustomerEvent;
import com.daiinfo.catering.util.ChangeDishEvent;
import com.daiinfo.catering.util.CustomerDaoFactory;
import com.daiinfo.catering.util.CustomerTableModel;
import com.daiinfo.catering.util.DishDaoFactory;
import com.daiinfo.catering.util.DishTableModel;
/*import product.ChangeEvent;
import product.DaoFactory;
import product.DaoInterface;
import product.ProductBean;
import product.ProductTableModel;
*/
public class DishManagePane2 extends JPanel {
    private JPanel panelTop = null;
    private JLabel labHeader = null;
    private JPanel panelBottom = null;
    private JButton add = null;
    private JButton delete = null;
    private JButton save = null;
    private JScrollPane scroll = null;
    private JTable table = null;
```

```
        private DishTableModel model = null;
        private List listDish = null;
        public DishTableModel getModel() {
            if (null == model) {
                model = new DishTableModel(listDish);
                // 给model添加一个监听，当修改的时候将触发该事件，代表事件的类是ChangeEvent
                model.addTableModelListener(new ChangeDishEvent(model));
                return model;
            }
            return model;
        }
        public JLabel getLabHeader() {
            if (null == labHeader) {
                labHeader = new JLabel("Maintaince Dish Information");
                return labHeader;
            }
            return labHeader;
        }
        public JTable getTable() {
            if (null == table) {
                table = new JTable(getModel());
                table.setEnabled(true);
                table.setRowSelectionAllowed(true);
                //table.setBackground(Color.blue);
                table.setSelectionForeground(Color.yellow);
                table.setSelectionBackground(Color.GRAY);
                //table.setForeground(Color.red);
                /**
                 * 隐藏第一列ID，不显示出来
                 */
                DefaultTableColumnModel dcm = (DefaultTableColumnModel) table.getColumnModel();
                dcm.getColumn(0).setMinWidth(0);
                dcm.getColumn(0).setMaxWidth(0);
                return table;
            }
            return table;
        }
        public JScrollPane getPanTable() {
            if (null == scroll) {
                scroll = new JScrollPane();
                scroll.setViewportView(getTable());
                return scroll;
            }
            return scroll;
        }
        public JPanel getPanelTop() {
            if (null == panelTop) {
                panelTop = new JPanel();
                panelTop.setLayout(new FlowLayout(FlowLayout.CENTER));
                panelTop.add(getLabHeader());
                return panelTop;
```

```
        }
        return panelTop;
    }
    public JPanel getPanelBottom() {
        if (null == panelBottom) {
            panelBottom = new JPanel();
            panelBottom.setLayout(new FlowLayout(FlowLayout.CENTER));
            panelBottom.add(getAdd());
            panelBottom.add(getDelete());
            panelBottom.add(getSave());
            return panelBottom;
        }
        return panelBottom;
    }
    public JButton getAdd() {
        /**
         * 点该按钮的时候调用addProduct()方法，在数据源（listProduct）将增加一个元素，没设值前都是null
         */
        if (null == add) {
            add = new JButton("添加");
            add.addActionListener(new ActionListener() {
                public void actionPerformed(ActionEvent e) {
                  DishesAddDialog dad=new DishesAddDialog();
                  dad.setVisible(true);
                }
            });
            return add;
        }
        return add;
    }
    public JButton getDelete() {
        if (null == delete) {
            delete = new JButton("删除");
            delete.addActionListener(new ActionListener() {
                public void actionPerformed(ActionEvent e) {
                    /**
                     * 支持一次选中多行后删除
                     */
                    int[] rows = getTable().getSelectedRows();
                    if (rows.length > 0) {
                        int flag = JOptionPane.showConfirmDialog(null, "确定删除？");
                        if (flag == JOptionPane.YES_OPTION)
                            deleteDish();
                    } else
                        JOptionPane.showMessageDialog(null, "请选择要删除的行！");
                }
            });
            return delete;
        }
        return delete;
    }
```

```
public JButton getSave() {
    if (null == save) {
        save = new JButton("保存");
        save.addActionListener(new ActionListener() {
            public void actionPerformed(ActionEvent e) {
                saveDish();
                JOptionPane.showMessageDialog(null, "更新成功！");
            }
        });
        return save;
    }
    return save;
}
public void addDish() {
    Dish dish = new Dish();
    getModel().addRow(getTable().getSelectedRow(), dish);
}
public void saveDish() {
    IBaseDao dao = DishDaoFactory.getDao();
    List changeList = getModel().getChangeList();
    // 如果有修改过就调用update方法
    if (changeList.size() > 0) {
        dao.update(changeList);
        changeList.clear();
    }
    List newRow = getModel().getNewRow();
    // 如果是新增就调用saveList，支持一次增加多行
    if (newRow.size() > 0) {
        dao.saveList(newRow);
        getModel().setList(dao.getList());
        getTable().updateUI();
        newRow.clear();
    }
}
public void deleteDish() {
    /**
     * 支持一次删除多行，先获得所有选中的行，然后按照行数取得Product实例，
     * 放进一个list，然后传给操作数据库的deleteList方法
     */
    int[] rows = getTable().getSelectedRows();
    ArrayList list = new ArrayList();
    IBaseDao dao = DishDaoFactory.getDao();
    for (int i = rows.length - 1; i >= 0; i--) {
        list.add(getModel().getRow(rows[i]));
        getModel().deleteRow(rows[i]);
    }
    dao.deleteList(list);
    getTable().updateUI();
    list.clear();
}
public void initData() {
```

```
        /**
         * 初始化数据源，从数据库里把数据拿出来，然后它会调用 getValueAt方法一个单元格一个单元格来设值，让它显示出来
         */
        listDish = new ArrayList();
        IBaseDao dao = DishDaoFactory.getDao();
        listDish = dao.getList();
    }
    public DishManagePane2() {
        initData();
        this.setLayout(new BorderLayout());
        add(getPanelTop(), BorderLayout.NORTH);
        add(getPanelBottom(), BorderLayout.SOUTH);
        add(getPanTable(), BorderLayout.CENTER);
        this.setSize(new Dimension(1024,800));
    }
}
```

3. 菜品实体类 Dish.java

```
package com.daiinfo.catering.entity;
import java.io.Serializable;
public class Dish implements Serializable {
    int id;
    String name;
    String code;
    String unit;
    double price;
    String status;
    Category category;
    String pic;
    public int getId() {
        return id;
    }
    public void setId(int id) {
        this.id = id;
    }
    public String getName() {
        return name;
    }
    public void setName(String name) {
        this.name = name;
    }
    public String getCode() {
        return code;
    }
    public void setCode(String code) {
        this.code = code;
    }
    public String getUnit() {
        return unit;
    }
```

```
    public void setUnit(String unit) {
        this.unit = unit;
    }
    public double getPrice() {
        return price;
    }
    public void setPrice(double price) {
        this.price = price;
    }
    public String getStatus() {
        return status;
    }
    public void setStatus(String status) {
        this.status = status;
    }
    public Category getCategory() {
        return category;
    }
    public void setCategory(Category category) {
        this.category = category;
    }

    public String getPic() {
        return pic;
    }
    public void setPic(String pic) {
        this.pic = pic;
    }
}
```

4. 菜品实现类 DishDaoImpl.java

```
package com.daiinfo.catering.dao;
import java.sql.Connection;
import java.sql.PreparedStatement;
import java.sql.ResultSet;
import java.sql.SQLException;
import java.util.ArrayList;
import java.util.Iterator;
import java.util.List;
import javax.swing.JOptionPane;
import com.daiinfo.catering.entity.Category;
import com.daiinfo.catering.entity.Customer;
import com.daiinfo.catering.entity.Dish;
import com.daiinfo.catering.util.Changed;
import com.daiinfo.catering.util.JDBConnection;
public class DishDaoImpl implements IBaseDao{
    Category category;
    CategoryDaoImpl cdi;

    public Dish getDishById(int id) {
```

```
        Dish dish = new Dish();
        Connection conn = JDBConnection.getConn();
        String s1 = "select * from dish where id=?";
        PreparedStatement ps = null;
        ResultSet rs = null;
        try {
            ps = conn.prepareStatement(s1);
            ps.setInt(1, id);
            rs = ps.executeQuery();
        } catch (SQLException e) {
            JOptionPane.showMessageDialog(null, "①取出dish全部数据出错! ");
            e.printStackTrace();
        }
        try {
            rs.next();
            dish.setId(rs.getInt("id"));

            int categoryid=rs.getInt("categoryId");
            cdi=new CategoryDaoImpl();
            category=new Category();
            category=cdi.getCategoryById(categoryid);
            dish.setCategory(category);

            dish.setName(rs.getString("name"));
            dish.setPic(rs.getString("pic"));
            dish.setCode(rs.getString("code"));
            dish.setUnit(rs.getString("unit"));
            dish.setPrice(rs.getDouble("price"));
            dish.setStatus(rs.getString("status"));
        } catch (SQLException e) {
            JOptionPane.showMessageDialog(null, "②取出dish数据出错! ");
            e.printStackTrace();
        } finally {
            try {
                rs.close();
                ps.close();
                conn.close();
            } catch (SQLException e) {
                JOptionPane.showMessageDialog(null, "关闭数据连接时出错! ");
                e.printStackTrace();
            }
        }
        return dish;
    }

    public Dish getDishByCode(String code) {
        Dish dish = new Dish();
        Connection conn = JDBConnection.getConn();
        String s1 = "select * from dish where code=?";
        PreparedStatement ps = null;
        ResultSet rs = null;
```

```
        try {
            ps = conn.prepareStatement(s1);
            ps.setString(1, code);
            rs = ps.executeQuery();
        } catch (SQLException e) {
            JOptionPane.showMessageDialog(null, "①取出dish全部数据出错！");
            e.printStackTrace();
        }
        try {
            rs.next();
            dish.setId(rs.getInt("id"));

            int categoryid=rs.getInt("categoryId");
            cdi=new CategoryDaoImpl();
            category=new Category();
            category=cdi.getCategoryById(categoryid);
            dish.setCategory(category);

            dish.setName(rs.getString("name"));
            dish.setPic(rs.getString("pic"));
            dish.setCode(rs.getString("code"));
            dish.setUnit(rs.getString("unit"));
            dish.setPrice(rs.getDouble("price"));
            dish.setStatus(rs.getString("status"));
        } catch (SQLException e) {
            JOptionPane.showMessageDialog(null, "②取出dish数据出错！");
            e.printStackTrace();
        } finally {
            try {
                rs.close();
                ps.close();
                conn.close();
            } catch (SQLException e) {
                JOptionPane.showMessageDialog(null, "关闭数据连接时出错！");
                e.printStackTrace();
            }
        }
        return dish;
    }

    @Override
    public List getList() {
        // TODO Auto-generated method stub
        Connection conn = JDBConnection.getConn();
        String s1 = "select * from dish order by id asc";
        PreparedStatement ps = null;
        ResultSet rs = null;
        List list = new ArrayList();

        category=new Category();
        cdi=new CategoryDaoImpl();
```

```
        try {
            ps = conn.prepareStatement(s1);
            rs = ps.executeQuery();
        } catch (SQLException e) {
            /*System.out.println("取出dish数据出错！");*/
            JOptionPane.showMessageDialog(null, "取出dish全部数据出错！");
            e.printStackTrace();
        }
        try {
            while (rs.next()) {
                Dish dish = new Dish();
                dish.setId(rs.getInt(1));
                category=cdi.getCategoryById(rs.getInt(2));
                dish.setCategory(category);
                dish.setName(rs.getString(3));
                dish.setPic(rs.getString(4));
                dish.setCode(rs.getString(5));
                dish.setUnit(rs.getString(6));
                dish.setPrice(rs.getDouble(7));
                dish.setStatus(rs.getString(8));
                list.add(dish);
            }
        } catch (SQLException e) {
            JOptionPane.showMessageDialog(null, "取出全部数据出错！");
            e.printStackTrace();
        } finally {
            try {
                rs.close();
                ps.close();
                conn.close();
            } catch (SQLException e) {
                JOptionPane.showMessageDialog(null, "关闭数据连接时出错！");
                e.printStackTrace();
            }
        }
        return list;
    }
    @Override
    public void saveList(List list) {
        // TODO Auto-generated method stub
        String s1 = "";
        Connection conn = JDBConnection.getConn();
        PreparedStatement ps = null;
        try {
            Iterator it = list.iterator();
            while (it.hasNext()) {
                Dish dish= (Dish) it.next();
                int id=dish.getId();
                int categoryId=dish.getCategory().getId();
                String name = dish.getName();
                String pic=dish.getPic();
```

```
                    String code=dish.getCode();
                    String unit=dish.getUnit();
                    double price=dish.getPrice();
                    String status=dish.getStatus();
                    s1 = "insert into dish(categoryId,name,pic,code,unit,price,status) values
(?,?,?,?,?,?,?)";
                    ps = conn.prepareStatement(s1);
                    ps.setInt(1, categoryId);
                    ps.setString(2, name);
                    ps.setString(3, pic);
                    ps.setString(4, code);
                    ps.setString(5, unit);
                    ps.setDouble(6, price);
                    ps.setString(7, status);
                    ps.executeUpdate();
                }
            } catch (SQLException e) {
                System.out.println("添加数据时出错！");
                JOptionPane.showMessageDialog(null, "添加数据时出错！");
                e.printStackTrace();
            } finally {
                try {
                    ps.close();
                    conn.close();
                } catch (SQLException e) {
                    JOptionPane.showMessageDialog(null, "关闭数据连接时出错！");
                    e.printStackTrace();
                }
            }
        }
        @Override
        public void deleteList(List list) {
            // TODO Auto-generated method stub
            String str = "delete from dish where id=?";
            int id = 0;
            Iterator it = list.iterator();
            Connection conn = JDBConnection.getConn();
            PreparedStatement ps = null;
            try {
                while (it.hasNext()) {
                    id = ((Dish) it.next()).getId();
                    ps = conn.prepareStatement(str);
                    ps.setInt(1, id);
                    ps.executeUpdate();
                }
            } catch (SQLException e) {
            } finally {
                try {
                    ps.close();
                    conn.close();
                } catch (SQLException e) {
```

```
                JOptionPane.showMessageDialog(null, "关闭数据连接时出错！");
                e.printStackTrace();
            }
        }
    }
    @Override
    public void update(List list) {
        // TODO Auto-generated method stub
        Connection conn = JDBConnection.getConn();
        PreparedStatement ps = null;
        Iterator it = list.iterator();
        int id = 0;
        int col = 0;
        String value = "";
        String str = "";
        try {
            while (it.hasNext()) {
                Changed ch = (Changed) it.next();
                id = ch.getId();
                col = ch.getCol();
                value = ch.getValue();
                switch (col) {
                case 1:
                    str = "update dish set categoryId=? where id=?";
                    break;
                case 2:
                    str = "update dish set name=? where id=?";
                    break;
                case 3:
                    str = "update dish set pic=? where id=?";
                    break;
                case 4:
                    str = "update dish set code=? where id=?";
                    break;
                case 5:
                    str = "update dish set unit=? where id=?";
                    break;
                case 6:
                    str = "update dish set price=? where id=?";
                    break;
                case 7:
                    str = "update dish set status=? where id=?";
                    break;
                }
                ps = conn.prepareStatement(str);
                ps.setString(1, value);
                ps.setInt(2, id);
                ps.executeUpdate();
            }
        } catch (SQLException e) {
            JOptionPane.showMessageDialog(null, "修改数据时出错！");
```

```
                e.printStackTrace();
            } finally {
                try {
                    ps.close();
                    conn.close();
                } catch (SQLException e) {
                    JOptionPane.showMessageDialog(null, "关闭数据连接时出错！");
                    e.printStackTrace();
                }
            }
        }

        public boolean save(Dish dish){
            Connection conn = JDBConnection.getConn();
            PreparedStatement ps = null;
            int categoryId=dish.getCategory().getId();
            String name = dish.getName();
            String pic=dish.getPic();
            String code=dish.getCode();
            String unit=dish.getUnit();
            double price=dish.getPrice();
            String status=dish.getStatus();
            String s1 = "insert into dish(categoryId,name,pic,code,unit,price,status) values
(?,?,?,?,?,?,?)";
            try {
                ps = conn.prepareStatement(s1);
                ps.setInt(1, categoryId);
                ps.setString(2, name);
                ps.setString(3, pic);
                ps.setString(4, code);
                ps.setString(5, unit);
                ps.setDouble(6, price);
                ps.setString(7, status);
                ps.executeUpdate();
            } catch (SQLException e) {
                // TODO Auto-generated catch block
                e.printStackTrace();
            }
            return true;
        }
    }
```

5. 表格事件处理类 ChangeDishEvent.java

```
    package com.daiinfo.catering.util;
    import java.util.ArrayList;
    import java.util.List;
    import javax.swing.event.TableModelEvent;
    import javax.swing.event.TableModelListener;
    public class ChangeDishEvent implements TableModelListener {
        /**
```

```
        */
        /**
         * 监听table被改动的事件，主要目的是用来记录被修改过的值，这样做可以一次任意行地修改值，修改一个单元格的值就记录一次
         * 主要记录id、新值、列数
         */
        DishTableModel model = null;
        public ChangeDishEvent(DishTableModel model) {
            this.model = model;
        }
        List list = model.getChangeList();
        int id = 0;
        String value = "";
        public void tableChanged(TableModelEvent arg0) {
            int row = arg0.getFirstRow();
            int col = arg0.getColumn();
            if (col != -1) {
                Changed cp = new Changed();
                id = ((Integer) model.getValueAt(row, 0)).intValue();
                if (id != 0) {
                    value = model.getValueAt(row, col).toString();
                    cp.setId(id);
                    cp.setCol(col);
                    cp.setValue(value);
                    list.add(cp);
                }
            }
        }
    }
```

6. 菜品表格类 DishTableModel.java

```
    package com.daiinfo.catering.util;
    import java.util.ArrayList;
    import java.util.Iterator;
    import java.util.List;
    import javax.swing.table.AbstractTableModel;
    import com.daiinfo.catering.dao.CategoryDaoImpl;
    import com.daiinfo.catering.entity.Category;
    import com.daiinfo.catering.entity.Customer;
    import com.daiinfo.catering.entity.Dish;
    public class DishTableModel extends AbstractTableModel {
        /**
         *
         * changeList用来存放被修改过的数据值，这样做是为了一次修改多行多值，保存的对象是ChangedProduct，只记录被修改过的值
         */
        private static List changeList = new ArrayList();
        private List list = new ArrayList();
        private String[] column = { "编号", "类别", "名称","图片","代码", "单位","单价","状态"};
        public DishTableModel() {
        }
```

```
    public DishTableModel(List list) {
        this();
        setList(list);
    }
    public int getColumnCount() {
        return column.length;
    }
    public int getRowCount() {
        return list.size();
    }
    /**
     * getValueAt方法就是使得数据在Table显示出来，给每个单元格设值
     */
    public Object getValueAt(int arg0, int arg1) {
        Dish dish = (Dish) list.get(arg0);
        return getPropertyValueByCol(dish, arg1);
    }
    public void addRow(int index, Dish dish) {
        if (index < 0 || index > list.size() - 1) {
            list.add(dish);
            fireTableRowsInserted(list.size(), list.size());
        } else {
            list.add(index + 1, dish);
            fireTableRowsInserted(index, index);
        }
    }
    public boolean deleteRow(int index) {
        if (index >= 0 && index < list.size()) {
            list.remove(index);
            fireTableRowsDeleted(index, index);
            return true;
        } else
            return false;
    }
    public boolean saveRow(int index,  Dish dish) {
        if (index >= 0 && index < list.size()) {
            list.set(index, dish);
            fireTableRowsUpdated(index, index);
            return true;
        } else
            return false;
    }
    public Dish getRow(int index) {
        if (index >= 0 && index < list.size()) {
            return (Dish) list.get(index);
        } else
            return null;
    }
    public List getNewRow() {
        List list = new ArrayList();
        List listDish = getList();
```

```
        Iterator it = listDish.iterator();
        while (it.hasNext()) {
            Dish dish = new Dish();
            dish = (Dish) it.next();
            if (dish.getId() == 0) {
                list.add(dish);
            }
        }
        return list;
    }
    public List getList() {
        return list;
    }
    public void setList(List list) {
        this.list = list;
        fireTableDataChanged();
    }
    public String getColumnName(int i) {
        return column[i];
    }
    public void setColumn(String[] column) {
        this.column = column;
    }
    public Object getPropertyValueByCol(Dish dish, int col) {
        switch (col) {
        case 0:
            return dish.getId();
        case 1:
            return dish.getCategory().getId();
        case 2:
            return dish.getName();
        case 3:
            return dish.getPic();
        case 4:
            return dish.getCode();
        case 5:
            return dish.getUnit();
        case 6:
            return dish.getPrice();
        case 7:
            return dish.getStatus();
        }
        return null;
    }
    public void setPropertyValueByCol(Dish dish, String value, int col) {
        switch (col) {
        case 1:
            CategoryDaoImpl cdi=new CategoryDaoImpl();
            Category category=new Category();
            category=cdi.getCategoryById(Integer.parseInt(value));
            dish.setCategory(category);;
```

```
                break;
            case 2:
                dish.setName(value);
                break;
            case 3:
                dish.setPic(value);
                break;
            case 4:
                dish.setCode(value);
                break;
            case 5:
                dish.setUnit(value);
                break;
            case 6:
                dish.setPrice(Integer.parseInt(value));
                break;
            case 7:
                dish.setStatus(value);
            }
            fireTableDataChanged();
        }
        public boolean isCellEditable(int row, int column) {
            return true;
        }
        /**
          * setValueAt方法是使增加或修改值的时候生效，aValue就是你在单元格填的值，要把这些值保存到数
据源中
         */
        public void setValueAt(Object aValue, int rowIndex, int columnIndex) {
            Dish dish = (Dish) list.get(rowIndex);
            setPropertyValueByCol(dish, aValue.toString(), columnIndex);
            this.fireTableCellUpdated(rowIndex, columnIndex);
        }
        public static List getChangeList() {
            return changeList;
        }
        public static void setChangeList(List changeList) {
            DishTableModel.changeList = changeList;
        }
    }
```

7. 工厂类 DishDaoFactory.java

```
    package com.daiinfo.catering.util;
    import com.daiinfo.catering.dao.CategoryDaoImpl;
    import com.daiinfo.catering.dao.CustomerDaoImpl;
    import com.daiinfo.catering.dao.DishDaoImpl;
    import com.daiinfo.catering.dao.IBaseDao;
    public class DishDaoFactory {
        /**
         *
         */
```

```
    synchronized public static IBaseDao getDao() {
        IBaseDao dao = null;
        if (dao == null) {
            dao = new DishDaoImpl();
            return dao;
        }
        return dao;
    }
}
```

8. 菜品添加界面类 DishesAddDialog.java

```
package com.daiinfo.category.dialog;
import java.awt.BorderLayout;
import java.awt.Component;
import java.awt.GridBagConstraints;
import java.awt.GridBagLayout;
import java.awt.GridLayout;
import java.awt.event.ActionEvent;
import java.awt.event.ActionListener;
import java.util.List;
import javax.swing.JButton;
import javax.swing.JComboBox;
import javax.swing.JDialog;
import javax.swing.JFrame;
import javax.swing.JLabel;
import javax.swing.JPanel;
import javax.swing.JTextField;
import com.daiinfo.catering.dao.CategoryDaoImpl;
import com.daiinfo.catering.dao.DishDaoImpl;
import com.daiinfo.catering.entity.Category;
import com.daiinfo.catering.entity.Dish;
public class DishesAddDialog extends JDialog {
    Dish dish = null;
    DishDaoImpl ddi = null;
    CategoryDaoImpl cdi = null;
    List<Category> listCategory = null;
    public DishesAddDialog() {
        super();
        setModal(true);
        setTitle("菜品管理");
        setBounds(100, 100, 500, 475);
        final JPanel inputPanel = new JPanel();
        inputPanel.setLayout(null);
        getContentPane().add(inputPanel, BorderLayout.CENTER);
        dish = new Dish();
        ddi = new DishDaoImpl();
        cdi = new CategoryDaoImpl();
        listCategory = cdi.getList();
        JLabel nameLabel = new JLabel("菜品名：");
        JTextField nameTxt = new JTextField(20);
        JLabel categoryIDLabel = new JLabel("菜品类别：");
```

```
JComboBox categoryIDCombo = new JComboBox();
for(Category cate:listCategory){
    categoryIDCombo.addItem(cate.getName());
}
JLabel picLabel = new JLabel("图片: ");
JTextField picTxt = new JTextField(10);
JLabel codeLabel = new JLabel("代码: ");
JTextField codeTxt = new JTextField(50);
JLabel unitLabel = new JLabel("单位: ");
JTextField unitTxt = new JTextField(20);
JLabel priceLabel = new JLabel("单价: ");
JTextField priceTxt = new JTextField(20);
JLabel statusLabel = new JLabel("状态: ");
JTextField statusTxt = new JTextField(20);
inputPanel.add(nameLabel);
inputPanel.add(nameTxt);
inputPanel.add(categoryIDLabel);
inputPanel.add(categoryIDCombo);
inputPanel.add(picLabel);
inputPanel.add(picTxt);
inputPanel.add(codeLabel);
inputPanel.add(codeTxt);
inputPanel.add(unitLabel);
inputPanel.add(unitTxt);
inputPanel.add(priceLabel);
inputPanel.add(priceTxt);
inputPanel.add(statusLabel);
inputPanel.add(statusTxt);
nameLabel.setBounds(120, 20, 160, 30);
nameTxt.setBounds(180, 20, 160, 30);
categoryIDLabel.setBounds(120, 60, 160, 30);
categoryIDCombo.setBounds(180, 60, 160, 30);
picLabel.setBounds(120, 100, 160, 30);
picTxt.setBounds(180, 100, 160, 30);
codeLabel.setBounds(120, 140, 160, 30);
codeTxt.setBounds(180, 140, 160, 30);
unitLabel.setBounds(120, 180, 160, 30);
unitTxt.setBounds(180, 180, 160, 30);
priceLabel.setBounds(120, 220, 160, 30);
priceTxt.setBounds(180, 220, 160, 30);
statusLabel.setBounds(120, 260, 160, 30);
statusTxt.setBounds(180, 260, 160, 30);
final JPanel buttonPanel = new JPanel();
buttonPanel.setLayout(new GridBagLayout());
getContentPane().add(buttonPanel, BorderLayout.SOUTH);
JButton okBtn = new JButton("确定");
JButton cancleBtn = new JButton("取消");
buttonPanel.add(okBtn);
buttonPanel.add(cancleBtn);
okBtn.addActionListener(new ActionListener(){
    @Override
```

```
        public void actionPerformed(ActionEvent e) {
            // TODO Auto-generated method stub
            Dish dish=new Dish();
            DishDaoImpl ddi=new DishDaoImpl();
            Category category=new Category();
            CategoryDaoImpl cdi=new CategoryDaoImpl();
            String str=(String) categoryIDCombo.getSelectedItem();
            category=cdi.getCategoryByName(str);

            dish.setCategory(category);
            dish.setName(nameTxt.getText());
            dish.setPic(picTxt.getText());
            dish.setCode(codeTxt.getText());
            dish.setUnit(unitTxt.getText());
            dish.setPrice(Integer.parseInt(priceTxt.getText()));
            dish.setStatus(statusTxt.getText());
            ddi.save(dish);
            setVisible(false);
        }

    });
    cancleBtn.addActionListener(new ActionListener(){
        @Override
        public void actionPerformed(ActionEvent e) {
            // TODO Auto-generated method stub
            setVisible(false);
        }
    });

  }
}
```

附录

附录一　Java 语言编码规范

Document number 文档编号	Confidentiality level 密级
	内部公开
Document version 文档版本	Total 227 pages 共 227 页
V1.00	

Java 语言编码规范

Prepared by 拟制		Date 日期	yyyy-mm-dd
Reviewed by 评审人		Date 日期	yyyy-mm-dd
Approved by 批准		Date 日期	yyyy-mm-dd

Revision Record 修订记录

Date 日期	Revision Version 修订版本	Sec No. 修改章节	Change Description 修改描述	Author 作者
yyyy-mm-dd	Vx.xx			

Table of Contents 目录

1. 范围

本规范规定了使用 Java 语言编程时排版、注释、命名、编码和 JTEST 的规则和建议。

本规范适用于使用 Java 语言编程的产品和项目。

2. 规范性引用文件

下列文件中的条款通过本规范的引用而成为本规范的条款。凡是注日期的引用文件，其随后所有的修改单（不包括勘误的内容）或修订版均不适用于本规范，然而，鼓励根据本规范达成协议的各方研究是否可使用这些文件的最新版本。凡是不注日期的引用文件，其最新版本适用于本规范。

序号	编号	名称
1	公司－DKBA1040-2001.12	《Java 语言编程规范》

3. 术语和定义

规则：编程时强制必须遵守的原则。

建议：编程时必须加以考虑的原则。

格式：对此规范格式的说明。

说明：对此规范或建议进行必要的解释。

示例：对此规范或建议从正、反两个方面给出例子。

4. 排版规范

4.1 规则

*4.1.1　程序块要采用缩进风格编写，缩进的空格数为 4 个。

说明：对于由开发工具自动生成的代码可以有不一致。

*4.1.2　分界符（如大括号 { 和 }）应各独占一行并且位于同一列，同时与引用它们的语句左对齐。在函数体的开始、类和接口的定义，以及 if、for、do、while、switch、case 语句中的程序都要采用如上的缩进方式。

示例：如下例子不符合规范。

```
for (…) {
    … // program code
}
if (…)
    {
    … // program code
    }

void example_fun( void )
    {
    … // program code
    }
```

应如下书写：

```
for (…)
{
    … // program code
}
if (…)
{
    … // program code
}

void example_fun( void )
{
    … // program code
}
```

*4.1.3　较长的语句、表达式或参数（>80 字符）要分成多行书写，长表达式要在低优先级操作符处划分新行，操作符放在新行之首，划分出的新行要进行适当的缩进，使排版整齐，语句可读。

示例：

```
if (filename != null
    && new File(logPath + filename).length() < LogConfig.getFileSize())
{
    … // program code
}

public static LogIterator read(String logType, Date startTime, Date endTime,
                               int logLevel, String userName, int bufferNum)
```

*4.1.4 不允许把多个短语句写在一行中，即一行只写一条语句

示例：如下例子不符合规范。

```
LogFilename now = null;         LogFilename that = null;
```

应如下书写：

```
LogFilename now = null;
LogFilename that = null;
```

*4.1.5 if、for、do、while、case、switch、default 等语句自占一行，且 if、for、do、while 等语句的执行语句无论多少都要加括号 {}。

示例：如下例子不符合规范。

```
if(writeToFile)           writeFileThread.interrupt();
```

应如下书写：

```
if(writeToFile)
{
    writeFileThread.interrupt();
}
```

*4.1.6 相对独立的程序块之间、变量说明之后必须加空行。

示例：如下例子不符合规范。

```
if(log.getLevel() < LogConfig.getRecordLevel())
{
    return;
}
LogWriter writer;
```

应如下书写：

```
if(log.getLevel() < LogConfig.getRecordLevel())
{
    return;
}
LogWriter writer;
int index;
```

*4.1.7 对齐只使用空格键，不使用 Tab 键。

说明：以免用不同的编辑器阅读程序时，因 Tab 键所设置的空格数目不同而造成程序布局不

整齐。JBuilder、UltraEdit 等编辑环境，支持行首 TAB 替换成空格，应将该选项打开。

*4.1.8　对两个以上的关键字、变量、常量进行对等操作时，它们之间的操作符之前、之后或者前后要加空格；进行非对等操作时，如果是关系密切的立即操作符（如 .），后不应加空格。

说明：采用这种松散方式编写代码的目的是使代码更加清晰。

由于留空格所产生的清晰性是相对的，所以，在已经非常清晰的语句中没有必要再留空格，如果语句已足够清晰，则括号内侧（即左括号后面和右括号前面）不需要加空格。多重括号间不必加空格，因为在 Java 语言中括号已经是最清晰的标志了。

在长语句中，如果需要加的空格非常多，那么应该保持整体清晰，而在局部不加空格。给操作符留空格时不要连续留两个以上空格。

示例：

（1）逗号、分号只在后面加空格。

```
int a, b, c;
```

（2）比较操作符，赋值操作符“=”“+=”，算术操作符“+”“%”，逻辑操作符“&&”“&”，位域操作符“<<”“^”等双目操作符的前后加空格。

```
if (current_time >= MAX_TIME_VALUE)
a = b + c;
a *= 2;
a = b ^ 2;
```

（3）“!”“~”“++”“--”“&”（地址运算符）等单目操作符前后不加空格。

```
flag = !isEmpty; // 非操作“!”与内容之间不加空格
i++;             // “++”,“--”与内容之间不加空格
```

（4）“.”前后不加空格。

```
p.id = pid;     // “.”前后不加空格
```

（5）if、for、while、switch 等与后面的括号间应加空格，使 if 等关键字更为突出、明显。

```
if (a >= b && c > d)
```

4.2　建议

类属性和类方法不要交叉放置，不同存取范围的属性或者方法也尽量不要交叉放置。

格式：

```
类定义
{
    类的公有属性定义
    类的保护属性定义
    类的私有属性定义
    类的公有方法定义
    类的保护方法定义
    类的私有方法定义
}
```

5. 注释规范

5.1 规则

5.1.1 一般情况下，源程序有效注释量必须在30%以上。

说明：注释的原则是有助于对程序的阅读理解，在该加的地方都加了，注释不宜太多也不能太少，注释语言必须准确、易懂、简洁。可以用注释统计工具来统计。

5.1.2 包的注释：写入一个名为 package.html 的 HTML 格式说明文件放入当前路径，作为包注释。

说明：方便 JavaDoc 收集。

示例：

```
com/huawei/msg/relay/comm/package.html
```

5.1.3 包的注释内容：简述本包的作用，详细描述本包的内容、产品模块名称和版本、公司版权。

说明：在详细描述中应该说明这个包的作用以及在整个项目中的位置。

格式：

```
<html>
<body>
<p>一句话简述
<p>详细描述
<p>产品模块名称和版本
<br>公司版权信息
</body>
</html>
```

示例：

```
<html>
<body>
<P>为 Relay 提供通信类，上层业务使用本包的通信类与SP进行通信
<p>详细描述
<p>MMSC V100R002 Relay
<br>(C) 版权所有 2012-2019 文思创新技术有限公司
</body>
</html>
```

5.1.4 文件注释：文件注释写入文件头部、包名之前的位置。

说明：注意以 /* 开始，避免被 JavaDoc 收集。

示例：

```
/*
 * 注释内容
 */
package com.huawei.msg.relay.comm;
```

5.1.5　文件注释内容：版权说明、描述信息、生成日期、修改历史。

说明：文件名可选。

格式：

```
/*
 * 文件名：[文件名]
 * 版权：〈版权〉
 * 描述：〈描述〉
 * 修改人：〈修改人〉
 * 修改时间：YYYY-MM-DD
 * 修改单号：〈修改单号〉
 * 修改内容：〈修改内容〉
 */
```

说明：每次修改后在文件头部写明修改信息，CheckIn 的时候可以直接把蓝色字体信息粘贴到 VSS 的注释上。在代码受控之前可以免去。

示例：

```
/*
 * 文件名：LogManager.java
 * 版权：Copyright 2012-2019 Huawei Tech. Co. Ltd. All Rights Reserved.
 * 描述： MMSC V100R002 Relay 通用日志系统
 * 修改人： 张三
 * 修改时间：2018-02-16
 * 修改内容：新增
 * 修改人： 李四
 * 修改时间：2018-02-26
 * 修改单号：WSS368
 * 修改内容：……
 * 修改人： 王五
 * 修改时间：2018-03-25
 * 修改单号：WSS498
 * 修改内容：……
 */
```

5.1.6　类和接口的注释：该注释放在 package 关键字之后，class 或者 interface 关键字之前。

说明：方便 JavaDoc 收集。

示例：

```
package com.huawei.msg.relay.comm;
/**
 * 注释内容
 */
public class CommManager
```

5.1.7　类和接口的注释内容：类的注释主要是一句话功能简述、功能详细描述。

说明：可根据需要列出版本号、生成日期、作者、内容、功能、与其他类的关系等。如果一个类存在 BUG（漏洞），请如实说明这些 BUG。

格式：

```
/**
 * 〈一句话功能简述〉
 * 〈功能详细描述〉
 * @author    [作者]
 * @version   [版本号, YYYY-MM-DD]
 * @see       [相关类/方法]
 * @since     [产品/模块版本]
 * @deprecated
 */
```

说明：描述部分说明该类或者接口的功能、作用、使用方法和注意事项，每次修改后增加作者和更新版本号和日期，@since 表示从此版本开始就有这个类或者接口，@deprecated 表示不建议使用该类或者接口。

示例：

```
/**
 * LogManager 类集中控制对日志读写的操作
 * 全部为静态变量和静态方法，对外提供统一接口。分配对应日志类型的读写器，
 * 读取或写入符合条件的日志记录
 * @author      张三，李四，王五
 * @version     1.2, 2018-03-25
 * @see          LogIteraotor
 * @see          BasicLog
 * @since        CommonLog 1.0
 */
```

5.1.8　类属性、公有和保护方法注释：写在类属性、公有和保护方法上面。

示例：

```
/**
 * 注释内容
 */
private String logType;
/**
 * 注释内容
 */
public void write()
```

5.1.9　成员变量注释内容：成员变量的意义、目的、功能，可能被用到的地方。

5.1.10　公有和保护方法注释内容：列出方法的一句话功能简述、功能详细描述、输入参数、输出参数、返回值、违例等。

格式：

```
/**
 * 〈一句话功能简述〉
 * 〈功能详细描述〉
 * @param  [参数1]    [参数1说明]
 * @param  [参数2]    [参数2说明]
 * @return [返回类型说明]
```

```
 * @exception/throws [违例类型] [违例说明]
 * @see           [类、类#方法、类#成员]
 * @deprecated
 */
```

说明：@since 表示从此版本开始就有这个方法；@exception 或 throws 列出可能仍出的异常；@deprecated 表示不建议使用该方法。

示例：

```
/**
 * 根据日志类型和时间读取日志。
 * 分配对应日志类型的LogReader，指定类型、查询时间段、条件和反复器缓冲数
 * 读取日志记录。查询条件为null或0表示无限制，反复器缓冲数为0读不到日志
 * 查询时间为左包含原则，即 [startTime, endTime)
 * @param logTypeName  日志类型名（在配置文件中定义的）
 * @param startTime    查询日志的开始时间
 * @param endTime      查询日志的结束时间
 * @param logLevel     查询日志的级别
 * @param userName     查询该用户的日志
 * @param bufferNum    日志反复器缓冲记录数
 * @return  结果集，日志反复器
 * @since  CommonLog1.0
 */
public static LogIterator read(String logType, Date startTime, Date endTime,
                               int logLevel, String userName, int bufferNum)
```

5.1.11　对于方法内部用 throw 语句抛出的异常，必须在方法的注释中标明，对于所调用的其他方法所抛出的异常，选择主要的在注释中说明。对于非 RuntimeException，即 throws 子句声明会抛出的异常，必须在方法的注释中标明。

说明：异常注释用 @exception 或 @throws 表示，在 JavaDoc 中两者等价，但推荐用 @exception 标注 Runtime 异常，@throws 标注非 Runtime 异常。异常的注释必须说明该异常的含义及什么条件下抛出该异常。

*5.1.12　注释应与其描述的代码相近，对代码的注释应放在其上方或右方（对单条语句的注释）相邻位置，不可放在下面，如放于上方则需与其上面的代码用空行隔开。

*5.1.13　注释与所描述内容进行同样的缩排。

说明：可使程序排版整齐，并方便注释的阅读与理解。

示例：如下例子，排版不整齐，阅读稍感不方便。

```
public void example( )
{
// 注释
    CodeBlock One

        // 注释
    CodeBlock Two
}
```

应改为如下布局：

```
public vid example()
{
    //注释
    CodeBlock one
    //注释
    CodeBlock two
}
```

*5.1.14　将注释与其上面的代码用空行隔开。

示例：如下例子，显得代码过于紧凑。

```
//注释
program code one
//注释
program code two
```

应如下书写：

```
//注释
program code one

//注释
program code two
```

*5.1.15　对变量的定义和分支语句（条件分支、循环语句等）必须编写注释。

说明：这些语句往往是程序实现某一特定功能的关键，对于维护人员来说，良好的注释可帮助其更好地理解程序，有时甚至优于看设计文档。

*5.1.16　对于 switch 语句下的 case 语句，如果因为特殊情况需要处理完一个 case 后进入下一个 case 处理，必须在该 case 语句处理完、下一个 case 语句前加上明确的注释。

说明：这样比较清楚程序编写者的意图，有效防止无故遗漏 break 语句。

*5.1.17　边写代码边注释，修改代码的同时修改相应的注释，以保证注释与代码的一致性。不再有用的注释要删除。

*5.1.18　注释的内容要清楚、明了，含义准确，防止注释二义性。

说明：错误的注释不但无益反而有害。

*5.1.19　避免在注释中使用缩写，特别是不常用缩写。

说明：在使用缩写时或之前，应对缩写进行必要的说明。

5.2　建议

*5.2.1　避免在一行代码或表达式的中间插入注释。

说明：除非必要，不应在代码或表达中间插入注释，否则容易使代码可理解性变差。

*5.2.2　通过对函数或过程、变量、结构等正确地命名以及合理地组织代码的结构，使代码成为自注释的。

说明：清晰准确的函数、变量等的命名，可增加代码的可读性，并减少不必要的注释。

*5.2.3　在代码的功能、意图层次上进行注释，提供有用、额外的信息。

说明：注释的目的是解释代码的目的、功能和采用的方法，提供代码以外的信息，帮助读者理解代码，防止没必要的重复注释信息。

示例：如下注释意义不大。

```
// 如果 receiveFlag 为真
if (receiveFlag)
```

而如下的注释则给出了额外有用的信息。

```
// 如果从连结收到消息
if (receiveFlag)
```

*5.2.4　在程序块的结束行右方加注释标记，以表明某程序块的结束。

说明：当代码段较长，特别是多重嵌套时，这样做可以使代码更清晰，更便于阅读。

示例：参见如下例子。

```
if (…)
{
    program code1

    while (index < MAX_INDEX)
    {
        program code2
    } // end of while (index < MAX_INDEX) // 指明该条while语句结束
} // end of  if (…) // 指明是哪条if语句结束
```

*5.2.5　注释应考虑程序易读及外观排版的因素，使用的语言若是中、英兼有的，建议多使用中文，除非能用非常流利准确的英文表达。

说明：注释语言不统一，影响程序易读性和外观排版，出于维护的考虑，建议使用中文。

5.2.6　方法内的单行注释使用 //。

说明：调试程序的时候可以方便地使用 /* …*/ 注释掉一长段程序。

5.2.7　注释尽量使用中文注释和中文标点。方法和类描述的第一句话尽量使用简洁明了的语句概括一下功能，然后加句号。接下来的部分可以详细描述。

说明：JavaDoc 工具收集简介的时候会选取第一句话。

5.2.8　顺序实现流程的说明使用 1、2、3、4 在每个实现步骤部分的代码前面进行注释。

示例：如下是对设置属性的流程注释

```
        //① 判断输入参数是否有效。
        …
        // ②设置本地变量。
        …
```

5.2.9　一些复杂的代码需要说明。

示例：这里主要是对闰年算法的说明。

```
        //① 如果能被4整除，是闰年；
        //② 如果能被100整除，不是闰年；
        //③ 如果能被400整除，是闰年。
```

6. 命名规范

6.1 规则

6.1.1　包名采用域后缀倒置加上自定义的包名，使用小写字母。在部门内部应该规划好包名的范围，防止产生冲突。部门内部产品使用部门的名称加上模块名称。产品线的产品使用产品的名称加上模块的名称。

格式：

```
com.huawei.产品名.模块名称
com.huawei.部门名称. 项目名称
```

示例：

```
Relay模块包名   com.huawei.msg.relay
通用日志模块包名 com.huawei.msg.log
```

6.1.2　类名和接口使用类意义完整的英文描述：每个英文单词的首字母使用大写、其余字母使用小写的大小写混合法。

```
示例：OrderInformation, CustomerList, LogManager, LogConfig
```

6.1.3　方法名使用类意义完整的英文描述：第一个单词的字母使用小写、剩余单词首字母大写其余字母小写的大小写混合法。

示例：

```
private void calculateRate();
public void addNewOrder();
```

6.1.4　方法中，存取属性的方法采用 setter 和 getter 方法，动作方法采用动词和动宾结构。

格式：

```
get + 非布尔属性名()
is + 布尔属性名()
set + 属性名()
动词()
动词 + 宾语()
```

示例：

```
public String getType();
public boolean isFinished();
public void setVisible(boolean);
public void show();
public void addKeyListener(Listener);
```

6.1.5　属性名使用意义完整的英文描述：第一个单词的字母使用小写、剩余单词首字母大写其余字母小写的大小写混合法。属性名不能与方法名相同。

示例：

```
private customerName;
private orderNumber;
private smpSession;
```

6.1.6　常量名使用全大写的英文描述，英文单词之间用下划线分隔开，并且使用 final static 修饰。

示例：

```
public final static int MAX_VALUE = 1000;
public final static String DEFAULT_START_DATE = "2001-12-08";
```

6.1.7　属性名可以和公有方法参数相同，不能和局部变量相同，引用非静态成员变量时使用 this 引用，引用静态成员变量时使用类名引用。

示例：

```
public class Person
{
    private String name;
    private static List properties;

    public void setName (String name)
    {
        this.name = name;
    }

    public void setProperties (List properties)
    {
        Person.properties = properties;
    }
}
```

6.2　建议

6.2.1　常用组件类的命名以组件名加上组件类型名结尾。

示例：

```
Application 类型的，命名以App 结尾——MainApp
Frame 类型的，命名以Frame 结尾——TopoFrame
Panel 类型的，建议命名以Panel 结尾——CreateCircuitPanel
Bean 类型的，建议命名以Bean 结尾——DataAccessBean
EJB 类型的，建议命名以EJB 结尾——DBProxyEJB
Applet 类型的，建议命名以Applet 结尾——PictureShowApplet
```

6.2.2　如果函数名超过 15 个字母，可采用以去掉元音字母的方法或者以行业内约定俗成的缩写方式缩写函数名。

示例：

```
getCustomerInformation()  改为  getCustomerInfo()
```

6.2.3　准确地确定成员函数的存取控制符号，不是必须使用 public 属性的，请使用 protected，不是必须使用 protected 的，请使用 private。

示例：

```
protected void setUserName(), private void  calculateRate()
```

6.2.4　含有集合意义的属性命名，尽量包含其复数的意义。

示例：

```
customers,  orderItems
```

7. 编码规范

7.1 规则

*7.1.1　明确方法功能，精确（而不是近似）地实现方法设计。一个函数仅完成一件功能，即使简单功能也应该编写方法实现。

说明：虽然为仅用一两行就可完成的功能去编方法好像没有必要，但用方法可使功能明确化，增加程序可读性，亦可方便维护、测试。

7.1.2　应明确规定对接口方法参数的合法性检查应由方法的调用者负责还是由接口方法本身负责，缺省是由方法调用者负责。

说明：对于模块间接口方法的参数的合法性检查这一问题，往往有两个极端现象，即：要么是调用者和被调用者对参数均不做合法性检查，结果就遗漏了合法性检查这一必要的处理过程，造成问题隐患；要么就是调用者和被调用者均对参数进行合法性检查，这种情况虽不会造成问题，但产生了冗余代码，降低了效率。

7.1.3　明确类的功能，精确（而非近似）地实现类的设计。一个类仅实现一组相近的功能。

说明：划分类的时候，应该尽量把逻辑处理、数据和显示分离，实现类功能的单一性。

示例：

```
数据类不能包含数据处理的逻辑。
通信类不能包含显示处理的逻辑。
```

7.1.4　所有的数据类必须重载 toString() 方法，返回该类有意义的内容。

说明：父类如果实现了比较合理的 toString()，子类可以继承不必再重写。

示例：

```
public TopoNode
{
    private String nodeName;
    public String toString()
    {
        return "NodeName : " + nodeName;
    }
}
```

7.1.5　数据库操作、IO 操作等需要使用结束 close() 的对象必须在 try -catch-finally 的 finally 中 close()。

示例：

```
try
{
```

```
        //……
    }
    catch(IOException ioe)
    {
        //……
    }
    finally
    {
        try
        {
            out.close();
        }
        catch (IOException ioe)
        {
            //……
        }
    }
```

7.1.6　异常捕获后，如果不对该异常进行处理，则应该记录日志或者 ex.printStackTrace() 。

说明：若有特殊原因，必须用注释加以说明。

示例：

```
try
{
    //……
}
catch (IOException ioe)
{
    ioe.printStackTrace ();
}
```

7.1.7　自己抛出的异常必须要填写详细的描述信息。

说明：便于问题定位。

示例：

```
throw new IOException("Writing data error! Data: " + data.toString());
```

7.1.8　运行期异常使用 RuntimeException 的子类来表示，不用在可能抛出异常的方法声明上加 throws 子句。非运行期异常是从 Exception 继承而来，必须在方法声明上加 throws 子句。

说明：

非运行期异常是由外界运行环境决定异常抛出条件的异常，例如文件操作，可能受权限、磁盘空间大小的影响而失败，这种异常是程序本身无法避免的，需要调用者明确考虑该异常出现时该如何处理方法，因此非运行期异常必须由 throws 子句标出，不标出或者调用者不捕获该类型异常都会导致编译失败，从而防止程序员本身疏忽。

运行期异常是程序在运行过程中本身考虑不周导致的异常，例如传入错误的参数等。抛出运行期异常的目的是防止异常扩散，导致定位困难。因此在做异常体系设计时要根据错误的性质合理选择自定义异常的继承关系。

还有一种异常是 Error 继承而来的，这种异常由虚拟机自己维护，表示发生了致命错误，程序无法继续运行（例如内存不足）。我们自己的程序不应该捕获这种异常，并且也不应该创建该种类型的异常。

7.1.9　在程序中使用异常处理还是使用错误返回码处理，根据是否有利于程序结构来确定，并且异常和错误码不应该混合使用，推荐使用异常。

说明：

一个系统或者模块应该统一规划异常类型和返回码的含义。

但是不能用异常来作为一般流程处理的方式，不要过多地使用异常，异常的处理效率比条件分支低，而且异常的跳转流程难以预测。

*7.1.10　注意运算符的优先级，并用括号明确表达式的操作顺序，避免使用默认优先级。

说明：防止阅读程序时产生误解，防止因默认的优先级与设计思想不符而导致程序出错。

示例：

下列语句中的表达式：

```
word = (high << 8) | low          ①
if ((a | b) && (a & c))           ②
if ((a | b) < (c & d))            ③
```

如果书写为：

```
high << 8 | low
a | b && a & c
a | b < c & d
```

①和②虽然不会出错，但语句不易理解；③造成了判断条件出错。

*7.1.11　避免使用不易理解的数字，用有意义的标识来替代。涉及物理状态或者含有物理意义的常量，不应直接使用数字，必须用有意义的静态变量来代替。

示例：如下的程序可读性差。

```
if (state == 0)
{
    state = 1;
    …// program code
}
```

应改为如下形式：

```
private final static int TRUNK_IDLE = 0;
private final static int TRUNK_BUSY = 1;
private final static int TRUNK_UNKNOWN = -1;

if (state == TRUNK_IDLE)
{
    state = TRUNK_BUSY;
    … // program code
}
```

7.1.12　数组声明的时候使用 int[] index，而不要使用 int index[]。

说明：使用 int index[] 格式使程序的可读性较差。

示例：

如下程序可读性差：

```
public int getIndex()[]
{
    …
}
```

如下程序可读性好：

```
public int[] getIndex()
{
    …
}
```

7.1.13 调试代码的时候，不要使用 System.out 和 System.err 进行打印，应该使用一个包含统一开关的测试类进行统一打印。

说明：代码发布的时候可以统一关闭调试代码，定位问题的时候又可以打开开关。

7.1.14 用调测开关来切换软件的 DEBUG 版和正式版，而不要同时存在正式版本和 DEBUG 版本的不同源文件，以减少维护的难度。

7.2 建议

7.2.1 记录异常不要保存 exception.getMessage()，而要记录 exception.toString()。

示例：

```
.NullPointException抛出时常常描述为空，这样往往看不出是出了什么错。
```

7.2.2 一个方法不应抛出太多类型的异常。

说明：如果程序中需要分类处理，则将异常根据分类组织成继承关系。如果确实有很多异常类型，首先考虑用异常描述来区别，throws/exception 子句标明的异常最好不要超过 3 个。

7.2.3 异常捕获尽量不要直接用 catch (Exception ex)，应该把异常细分处理。

*7.2.4 如果多段代码重复做同一件事情，那么在方法的划分上可能存在问题。

说明：若此段代码各语句之间有实质性关联并且是完成同一件功能的，那么可考虑把此段代码构造成一个新的方法。

7.2.5 对于创建的主要的类，最好置入 main() 函数，包含用于测试那个类的代码。

说明：主要类包括以下几项。

（1）能完成独立功能的类，如通信。

（2）具有完整界面的类，如一个对话框、一个窗口、一个帧等。

（3）JavaBean 类。

示例：

```
public static void main(String[] arguments)
{
    CreateCircuitDialog circuitDialog1 = new CreateCircuitDialog (null,
                                                    "Ciruit",  false);
```

```
    circuitDialog1.setVisible(true);
}
```

7.2.6　集合中的数据如果不使用了应该及时释放，尤其是可重复使用的集合。

说明：由于集合保存了对象的句柄，虚拟机的垃圾收集器就不会回收。

*7.2.7　源程序中关系较为紧密的代码应尽可能相邻。

说明：便于程序阅读和查找。

示例：矩形的长与宽关系较密切，应放在一起。

```
rect.length = 10;
rect.width = 5;
```

*7.2.8　不要使用难懂的技巧性很高的语句，除非很有必要时。

说明：高技巧语句不等于高效率的程序，实际上程序的效率关键在于算法。

8. JTEST 规范

8.1　规则

1. 在 switch 中每个 case 语句都应该包含 break 或者 return。
2. 不要使用空的 for、if、while 语句。
3. 在运算中不要减小数据的精度。
4. switch 语句中的 case 关键字要和后面的常量保持一个空格，switch 语句中不要定义 case 之外的无用标签。
5. 不要在 if 语句中使用等号 = 进行赋值操作。
6. 静态成员或者方法使用类名访问，不使用句柄访问。
7. 方法重载的时候，一定要注意方法名相同，避免类中使用两个非常相似的方法名。
8. 不要在 ComponentListener.componentResized() 方法中调用 serResize() 方法。
9. 不要覆盖父类的静态方法和私有方法。
10. 不要覆盖父类的属性。
11. 不要使用两级以上的内部类。
12. 把内部类定义成私有类。
13. 去掉接口中多余的定义（不使用 public、abstract、static、final 等，这是接口中默认的）。
14. 不要定义不会被用到的局部变量、类私有属性、类私有方法和方法参数。
15. 显式初始化所有的静态属性。
16. 不要使用 System.getenv() 方法。
17. 不要硬编码“\n”和“\r”作为换行符号。
18. 不要直接使用 Java.awt.peer.* 里面的接口。
19. 使用 System.arraycopy()，不使用循环来复制数组。
20. 避免不必要的 instanceof 比较运算和类造型运算。
21. 不要在 finalize() 方法中删除监听器（Listeners）。
22. 在 finalize() 方法中一定要调用 super.finalize() 方法。

23. 在 finalize() 方法中的 finally 中调用 super.finalize() 方法。
24. 进行字符转换的时候应该尽可能地减少临时变量。
25. 使用 ObjectStream 的方法后，调用 reset()，释放对象。
26. 线程同步中，在循环里面使用条件测试（使用 while(isWait) wait() 代替 if(isWait) wait() ）。
27. 不掉用 Thread 类的 resume()、suspend()、stop() 方法。
28. 减小单个方法的复杂度，使用的 if、while、for、switch 语句要在 10 个以内。
29. 在 Servlets 中，重用 JDBC 连接的数据源。
30. 减少在 Sevlets 中使用的同步方法。
31. 不定义在包中没有被用到的友好属性、方法和类。
32. 没有子类的友好类应该定义成 final。
33. 没有被覆盖的友好方法应该定义成 final。

8.2 建议

1. 为 switch 语句提供一个 default 选项。
2. 不要在 for 循环体中对计数器赋值。
3. 不要给非公有类定义 public 构建器。
4. 不要对浮点数进行比较运算，尤其是不要进行 == 和 != 运算，减少 > 和 < 运算。
5. 实现 equals() 方法时，先用 getClass() 或 instanceof 进行类型比较，通过后才能继续比较。
6. 不要重载 main() 方法用作除入口以外的其他用途。
7. 方法的参数名不要和类中的方法名相同。
8. 除了构建器外，不要使用和类名相同的方法名。
9. 不要定义 Error 和 RuntimeException 的子类，可以定义 Exception 的子类。
10. 线程中需要实现 run() 方法。
11. 使用 equals() 比较两个类的值是否相同。
12. 字符串和数字运算结果相连接的时候，应该把数字运算部分用小括号括起来。
13. 类中不要使用非私有（公有、保护和友好）的非静态属性。
14. 在类中对于没有实现的接口，应该定义成抽象方法，类应该定义成抽象类（5 级）。
15. 不要显式导入 Java.lang.* 包。
16. 初始化时不要使用类的非静态属性。
17. 显式初始化所有的局部变量。
18. 按照方法名把方法排序放置，同名和同类型的方法应该放在一起。
19. 不要使用嵌套赋值，即在一个表达式中使用多个 =。
20. 不要在抽象类的构建器中调用抽象方法。
21. 重载 equals() 方法的同时，也应该重载 hashCode() 方法。
22. 工具类（Utility）不要定义构建器，包括私有构建器。
23. 不要在 switch 中使用 10 个以上的 case 语句。
24. 把 main() 方法放在类的最后。
25. 声明方法违例的时候不要使用 Exception，应该使用它的子类。
26. 不要直接扔出一个 Error，应该扔出它的子类。
27. 在进行比较的时候，总是把常量放在同一边（都放在左边或者都放在右边）。

28. 在可能的情况下，总是为类定义一个缺省的构建器。

29. 在捕获违例的时候，不使用 Exception、RuntimeException、Throwable，尽可能使用它们的子类。

30. 在接口或者工具类中定义常量（5 级）。

31. 使用大写“L”表示 long 常量（5 级）。

32. main() 方法必须是 public static void main(String[])（5 级）。

33. 对返回类型为 boolean 的方法使用 is 开头，其他类型的不能使用。

34. 对非 boolean 类型取值方法（getter）使用 get 开头，其他类型的不能使用。

35. 对于设置值的方法（setter）使用 set 开头，其他类型的不能使用。

36. 方法需要有同样数量参数的注释 @param。

37. 不要在注释中使用不支持的标记，如：@unsupported。

38. 不要使用 Runtime.exec() 方法。

39. 不要自定义本地方法（native method）。

40. 使用尽量简洁的运算符号。

41. 使用集合时设置初始容量。

42. 单个首字符的比较使用 charAt() 而不用 startsWith()。

43. 对于被除数或者被乘数为 2 的 *n* 次方的乘除运算使用移位运算符 >> 及 <<。

44. 一个字符的连接使用 ' ' 而不使用 " "，如：String a = b + 'c'。

45. 不要在循环体内调用同步方法和使用 try-catch 块。

46. 不要使用不必要的布尔值比较，如：if (a.equals(b))，而不是 if (a.equals(b)==true)。

47. 常量字符串使用 String，非常量字符串使用 StringBuffer。

48. 在循环条件判断的时候不要使用复杂的表达式。

49. 对于“if (condition) do1; else do2;”语句使用条件操作符“if (condition)?do1:do2;”。

50. 不要在循环体内定义变量。

51. 使用 StringBuffer 的时候设置初始容量。

52. 尽可能地使用局部变量进行运算。

53. 尽可能少地使用“!”操作符（5 级）。

54. 尽可能地对接口进行 instanceof 运算（5 级）。

55. 不要使用 Date[] 而要使用 long[] 替代。

56. 不要显式调用 finalize() 。

57. 不要使用静态集合，其内存占用增长没有边界。

58. 不要重复调用一个方法获取对象，使用局部变量重用对象。

59. 线程同步中，使用 notifyAll() 代替 notify()。

60. 避免在同步方法中调用另一个同步方法造成的死锁。

61. 非同步方法中不能调用 wait() 和 notify() 方法。

62. 使用 wait() 和 notify() 代替 while()、sleep()。

63. 不要使用同步方法，使用同步块（5 级）。

64. 把所有的公有方法定义为同步方法（5 级）。

65. 实现的 Runnable.run() 方法必须是同步方法（5 级）。

66. 一个只有 abstract 方法、final static 属性的类应该定义成接口。

67. 在 clone() 方法中应该而且必须使用 super.clone() 而不是 new。
68. 常量必须定义为 final。
69. 在 for 循环中提供终止条件。
70. 在 for，while 循环中使用增量计数。
71. 使用 StringTokenizer 代替 indexOf() 和 substring()。
72. 不要在构建器中使用非 final 方法。
73. 不要对参数进行赋值操作（5 级）。
74. 不要通过名字比较两个对象的类，应该使用 getClass()。
75. 安全：尽量不要使用内部类。
76. 安全：尽量不要使类可以克隆。
77. 安全：尽量不要使接口可以序列化。
78. 安全：尽量不要使用友好方法、属性和类。
79. Servlet：不要使用 Java.beans.Beans.instantiate() 方法。
80. Servlet：不再使用 HttpSession 时，应该尽早使用 invalidate() 方法释放。
81. Servlet：不再使用 JDBC 资源时，应该尽早使用 close() 方法释放。
82. Servlet：不要使用 Servlet 的 SingleThreadModel，会消耗大量资源。
83. 国际化：不要使用一个字符进行逻辑操作，使用 Characater。
84. 国际化：不要进行字符串连接操作，使用 MessageFormat。
85. 国际化：不要使用 Date.toString() 和 Time.toString() 方法。
86. 国际化：字符和字符串常量应该放在资源文件中。
87. 国际化：不要使用数字的 toString() 方法。
88. 国际化：不要使用 StringBuffer 和 StringTokenizer 类。
89. 国际化：不要使用 String 类的 compareTo() 及 equals() 方法。
90. 复杂度：建议的最大规模如下。

```
继承层次        5层
类的行数        1000行（包含{}）
类的属性        10个
类的方法        20个
类友好方法      10个
类私有方法      15个
类保护方法      10个
类公有方法      10个
类调用方法      20个
方法参数        5个
return语句      1个
方法行数        30行
方法代码        20行
注释比率        30%~50%
```

附录二　Java 注释模板设置

设置注释模板的入口：Window → Preference → Java → Code Style → Code Template，然后展开 Comments 节点就是所有需设置注释的元素了。现就每一个元素逐一介绍。

文件（Files）注释标签

```
/**
 * @Title: ${file_name}
 * @Package ${package_name}
 * @Description: ${todo}(用一句话描述该文件做什么)
 * @author JoJo
 * @date ${date} ${time}
 * @version V1.0
 */
```

类型（Types）注释标签（类的注释）

```
/**
 * @ClassName: ${type_name}
 * @Description: ${todo}(这里用一句话描述这个类的作用)
 * @author JoJo
 * @date ${date} ${time}
 *
 * ${tags}
 */
```

字段（Fields）注释标签

```
/**
 * @Fields ${field} : ${todo}(用一句话描述这个变量表示什么)
 */
```

构造函数（Constructor）标签

```
/**
 * <p>Title: </p>
 * <p>Description: </p>
 * ${tags}
 */
```

方法（Methods）标签

```
/**
 * @Title: ${enclosing_method}
 * @Description: ${todo}(这里用一句话描述这个方法的作用)
 * @param ${tags}     设定文件
 * @return ${return_type}     返回类型
 * @throws
 */
```

覆盖方法（Overriding Methods）标签

```
/* (非 javadoc)
 * <p>Title: ${enclosing_method}</p>
 * <p>Description: </p>
 * ${tags}
 * ${see_to_overridden}
 */
```

代表方法（Delegate Methods）标签

```
/**
 * ${tags}
 * ${see_to_target}
 */
```

getter 方法标签

```
/**
 * @return the ${bare_field_name}
 */
```

setter 方法标签

```
/**
 * @param ${param} the ${bare_field_name} to set
 */
```

附录三　常用 Java 正则表达式

一、校验数字的表达式

1. 数字：^[0-9]*$
2. n 位的数字：^\d{n}$
3. 至少 n 位的数字：^\d{n,}$
4. m ~ n 位的数字：^\d{m,n}$
5. 零和非零开头的数字：^(0|[1-9][0-9]*)$
6. 非零开头的最多带两位小数的数字：^([1-9][0-9]*)+(.[0-9]{1,2})?$
7. 带 1 ~ 2 位小数的正数或负数：^(\-)?\d+(\.\d{1,2})?$
8. 正数、负数和小数：^(\-|\+)?\d+(\.\d+)?$
9. 有两位小数的正实数：^[0-9]+(.[0-9]{2})?$
10. 有 1 ~ 3 位小数的正实数：^[0-9]+(.[0-9]{1,3})?$
11. 非零的正整数：^[1-9]\d*$ 或 ^([1-9][0-9]*){1,3}$ 或 ^\+?[1-9][0-9]*$
12. 非零的负整数：^\-[1-9][]0-9"*$ 或 ^-[1-9]\d*$
13. 非负整数：^\d+$ 或 ^[1-9]\d*|0$
14. 非正整数：^-[1-9]\d*|0$ 或 ^((-\d+)|(0+))$
15. 非负浮点数：^\d+(\.\d+)?$ 或 ^[1-9]\d*\.\d*|0\.\d*[1-9]\d*|0?\.0+|0$
16. 非正浮点数：^((-\d+(\.\d+)?)|(0+(\.0+)?))$ 或 ^(-([1-9]\d*\.\d*|0\.\d*[1-9]\d*))|0?\.0+|0$
17. 正浮点数：^[1-9]\d*\.\d*|0\.\d*[1-9]\d*$ 或 ^(([0-9]+\.[0-9]*[1-9][0-9]*)|([0-9]*[1-9][0-9]*\.[0-9]+)|([0-9]*[1-9][0-9]*))$
18. 负浮点数：^-([1-9]\d*\.\d*|0\.\d*[1-9]\d*)$ 或 ^(-(([0-9]+\.[0-9]*[1-9][0-9]*)|([0-9]*[1-9][0-9]*\.[0-9]+)|([0-9]*[1-9][0-9]*)))$
19. 浮点数：^(-?\d+)(\.\d+)?$ 或 ^-?([1-9]\d*\.\d*|0\.\d*[1-9]\d*|0?\.0+|0)$

二、校验字符的表达式

1. 汉字：^[\u4e00-\u9fa5]{0,}$
2. 英文和数字：^[A-Za-z0-9]+$ 或 ^[A-Za-z0-9]{4,40}$
3. 长度为 3 ~ 20 的所有字符：^.{3,20}$
4. 由 26 个英文字母组成的字符串：^[A-Za-z]+$
5. 由 26 个大写英文字母组成的字符串：^[A-Z]+$
6. 由 26 个小写英文字母组成的字符串：^[a-z]+$
7. 由数字和 26 个英文字母组成的字符串：^[A-Za-z0-9]+$
8. 由数字、26 个英文字母或者下划线组成的字符串：^\w+$ 或 ^\w{3,20}$
9. 中文、英文、数字包括下划线：^[\u4E00-\u9FA5A-Za-z0-9_]+$
10. 中文、英文、数字但不包括下划线等符号：^[\u4E00-\u9FA5A-Za-z0-9]+$ 或 ^[\u4E00-\u9FA5A-Za-z0-9]{2,20}$
11. 可以输入含有 ^%&',;=?$\" 等字符：[^%&',;=?$\x22]+

12. 禁止输入含有~的字符：[^~\x22]+

三、特殊需求表达式

1. Email 地址：^\w+([-+.]\w+)*@\w+([-.]\w+)*\.\w+([-.]\w+)*$

2. 域名：[a-zA-Z0-9][-a-zA-Z0-9]{0,62}(/.[a-zA-Z0-9][-a-zA-Z0-9]{0,62})+/.?

3. InternetURL：[a-zA-z]+://[^\s]* 或 ^http://([\w-]+\.)+[\w-]+(/[\w-./?%&=]*)?$

4. 手机号码：^(13[0-9]|14[5|7]|15[0|1|2|3|5|6|7|8|9]|18[0|1|2|3|5|6|7|8|9])\d{8}$

5. 电话号码（XXX-XXXXXXX、XXXX-XXXXXXXX、XXX-XXXXXXX、XXX-XXXXXXXX、XXXXXXX 和 XXXXXXXX）：^(\(\d{3,4}-)|\d{3.4}-)?\d{7,8}$

6. 国内电话号码（0511-4405222、021-87888822）：\d{3}-\d{8}|\d{4}-\d{7}

7. 身份证号（15 位、18 位数字）：^\d{15}|\d{18}$

8. 短身份证号码（数字、字母 x 结尾）：^([0-9]){7,18}(x|X)?$ 或 ^\d{8,18}|[0-9x]{8,18}|[0-9X]{8,18}?$

9. 账号是否合法（字母开头，允许 5 ~ 16 字节，允许字母、数字和下画线）：^[a-zA-Z][a-zA-Z0-9_]{4,15}$

10. 密码（以字母开头，长度在 6 ~ 18 之间，只能包含字母、数字和下画线）：^[a-zA-Z]\w{5,17}$

11. 强密码（必须包含大小写字母和数字的组合，不能使用特殊字符，长度在 8 ~ 10 之间）：^(?=.*\d)(?=.*[a-z])(?=.*[A-Z]).{8,10}$

12. 日期格式：^\d{4}-\d{1,2}-\d{1,2}

13. 一年的 12 个月（01 ~ 09 和 1 ~ 12）：^(0?[1-9]|1[0-2])$

14. 一个月的 31 天（01 ~ 09 和 1 ~ 31）：^((0?[1-9])|((1|2)[0-9])|30|31)$

15. 钱的输入格式如下。

① 有 4 种钱的表示形式我们可以接受：10000.00 和 10,000.00，和没有“分”的 10000 和 10,000：^[1-9][0-9]*$

② 这表示任意一个不以 0 开头的数字，但是，这也意味着一个字符“0”不通过，所以我们采用下面的形式：^(0|[1-9][0-9]*)$

③ 一个 0 或者一个不以 0 开头的数字。我们还可以允许开头有一个负号：^(0|-?[1-9][0-9]*)$

④ 这表示一个 0 或者一个可能为负的开头不为 0 的数字。让用户以 0 开头好了，把负号的也去掉，因为钱总不能是负的吧。下面我们要加的是说明可能的小数部分：^[0-9]+(.[0-9]+)?$

⑤ 必须说明的是，小数点后面至少应该有 1 位数，所以“10.”是不通过的，但是“10”和“10.2”是通过的：^[0-9]+(.[0-9]{2})?$

⑥ 这样我们规定小数点后面必须有两位，如果你认为太苛刻了，可以这样：^[0-9]+(.[0-9]{1,2})?$

⑦ 这样就允许用户只写一位小数。下面我们该考虑数字中的逗号了，我们可以这样：^[0-9]{1,3}(,[0-9]{3})*(.[0-9]{1,2})?$

⑧ 1 到 3 个数字，后面跟着任意个 逗号 +3 个数字，逗号成为可选，而不是必须：^([0-9]+|[0-9]{1,3}(,[0-9]{3})*)(.[0-9]{1,2})?$

备注：这就是最终结果了，别忘了“+”可以用“*”替代。如果你觉得空字符串也可以接受的话（奇怪，为什么？）最后，别忘了在用函数时去掉那个反斜杠，一般的错误都在这里。